Physics Without Energy Fields or Charge

Delta-C Mechanics

DT Froedge

Hardcover ISBN: 979-8-218-82971-1
Paperback ISBN:979-8-218-82972-8

Cover design, book design, and typesetting by Mayfly book design

Library of Congress Catalog Number: 2025922091

First Printing: 2025

Contents

This paper proposes the Feynman flow density as a new, physically intuitive framework for modeling the hydrogen atom and its fine structure. Instead of treating the electron as a point charge described by an abstract wavefunction, the approach represents it as a bound state of circulating photon probability flows that also extend outward to interact with the proton. By reformulating Feynman's path integral in terms of real action flows rather than phase amplitudes, the method recovers the entire hydrogen spectral series and fine-structure splittings with accuracy matching high-precision NIST data. The work aims both to reproduce quantum predictions without relying on the electric field concept and to restore a tangible mechanical model of hydrogen, updating the intuitive picture once offered by Bohr while remaining compatible with Schrödinger and Dirac theory.

Dedication

To my wife, Monica, for her patience with the time I've spent tinkering with "a little physics."

Once, when we were sitting in the backyard watching the sunset, I told her that I had found a most extraordinary discovery regarding the relation between electricity and gravitation. She said, "I hope that dog out there isn't crapping on the walkway."

DT Froedge

Academics

B.S. Physics, Mathematics. Western. Kentucky University, 1965
M.S Physics. University of Tennessee, 1967
PhD, Physics, Auburn University, Academic Course Completed, 1969

Professional

American Physical Society, 2005-Present
License Professional Engineer, PA, CT, NC TX, AL.
CEO, GeoSonics & Vibra-Tech Engineers Inc.
E-mail: DT.Froedge903@topper.wku.edu

Foreword

This work represents the author's second book on mechanics. The first volume served as both a historical archive and a developmental record of early theoretical progress, including exploratory paths, false starts, and conceptual missteps. While that material may primarily interest the author, it remains a valuable resource for understanding the evolution of the theory and the origins of its more unconventional ideas.

This second volume adopts a more focused and analytical approach, addressing foundational aspects of classical and modern mechanics—particularly phenomena such as centrifugal force and the Lorentz transformation—and exploring their deeper implications within an extended framework referred to as Delta-c Mechanics.

Of particular significance is the emerging connection between this theoretical framework and the formalism of Feynman's path integrals. In standard quantum mechanics, Feynman's action integral approach provides a powerful method for computing transition amplitudes by summing over all possible paths, weighted by the exponential of the classical action (in units of $\hbar$). The discovery that the action flow functions derived in Delta-c Mechanics closely mirror the mathematical structure of these path integrals suggests a deeper, perhaps fundamental, link between atomic state transitions and the mass properties of nuclear particles.

This consilience offers compelling evidence that the probability density flows, which govern quantum transitions, may also inform the structural basis of particle mass and vertex functions in interaction theory. The vertex functions—central to quantum field theory—represent interaction points where particles are created, annihilated, or scattered, and they play a key role in defining both atomic and subatomic dynamics.

The realization that flow probability densities in Delta-c Mechanics correspond closely with these standard constructions marks a noteworthy advance. It not only bridges concepts from classical mechanics to quantum field theory but also offers a novel interpretation of how atomic states may emerge from deeper nuclear and subnuclear principles.

The first two papers included in this volume are arguably the most significant in establishing the theoretical foundation. They define the core operations of the framework and articulate its relevance to both established physics and novel extensions.

Preface

This book represents the second volume in the author's theoretical exploration of the fundamental interaction between mass and electromagnetic energy. At the heart of this research lies the proposition that mass arises from the localization of photons—specifically through the self-binding of rotating photon pairs. This framework reimagines mass not as an intrinsic property, but as an emergent phenomenon grounded in electrodynamic structure.

A key premise of the theory is that the speed of light is not merely a fixed universal limit, but a result of deeper interactions—specifically, the opposing flow and scattering effects within a vacuum medium. This vacuum is modeled as a dynamic density field, shaped by the Feynman action flow density generated by the photons comprising all mass particles in the universe. Within this view, the vacuum is not empty but an active participant in energy-mass behavior.

The first book was a compilation of early-stage papers, capturing the development of these ideas from their inception. It began with a radical departure from conventional physics—intentionally setting aside traditional constructs such as force, gravitational potential, and even quantum wavefunctions. Instead, the approach focused on the use of fundamental constants alone, seeking to uncover what could be derived from first principles. Although this led to countless false starts, certain recurring relationships began to stand out—among the earliest being an intriguing link between the fine-structure constant (α) and the gravitational constant (G), hinting at a deeper unity between electromagnetic and gravitational phenomena.

The Feynman action integral, central to quantum electrodynamics, soon emerged as a promising formal tool. Yet it wasn't until more recently that its full relevance became apparent—particularly the interpretation of its flow density as distinct from the probabilistic interpretation of the wavefunction in standard quantum mechanics. This contrast opened new avenues for understanding the mechanisms by which particle states and masses arise.

This second volume is more focused and interpretive in nature. It aims to clarify how the Delta-c Mechanics framework intersects with and offers fresh insight into well-established but often poorly understood concepts in classical and relativistic mechanics—such as the Lorentz transformation, the origin of centrifugal force, the invariance of the speed of light, and the quantized values of particle mass and energy states.

This book is an ongoing development. The chronology goes from last to first, as the consilience of atomic and nuclear structure has been better understood.

Together, these two books represent the culmination of a lifelong quest to better understand the underlying structure of the physical world. The author's academic journey began at Western Kentucky University in 1961 and concluded with the completion of the academic requirements for a PhD in Physics at Auburn University in 1969. My final dissertation, though never formally submitted, focused on the interaction of shock waves in plasma.

This work is offered as another direction in the quest for an understanding of the mechanics of the universe. It is quite radical for its time and not likely to find its way in a world of strings, non-local interactions, dark energy, and mass. It is too right to be wrong, however—as some future generation will discover. ~DTF 6/14/2025

Feynman Flow Density Alternative to Wavefunctions

D.T. Froedge

V060225

$$\psi^{*}\psi \qquad \phi^{*}\phi$$

Preface

The QM wavefunction ψ is purely mathematical structure, not physically related to anything observable and the amplitude is not definable. It works, but no one is certain as to why. This paper will define a vector flow density of the Feynman action path integrals $\phi^{*}\phi$ as an alternate probability function to defined quantum states, and addresses some of the conceptual issues.

The Feynman action path integrals are generally assumed to be the amplitude of a wavefunction and, with the end point being the sum over all the paths, creating a vertex function. The phase summations create the state but have no effect on the phase or motion of other particles. These same action functions can be taken as the probable flow density of photons generated by the photon on action paths inside mass particles. The difference mathematically is slight, but the probable action flow creates the difference between a passive probability of state existence and the mechanism that interacts with the action flow of other particles to create state values. The flow interaction creates the effect of charge and energy fields without the necessity of postulating the existence of either.

Introduction

An electron is composed of two photons consisting of two rotating Planck particles revolving around their center of mass at the Compton radius

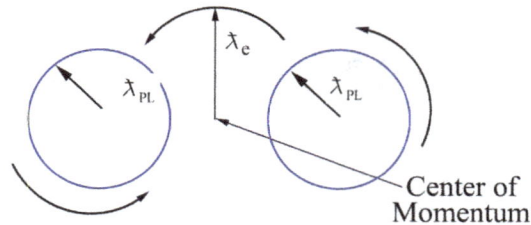

Fig. 1: The Electron

In earlier work regarding the vacuum flow density of space and the change in the velocity of light in regard to the flow density around the electron, it was discovered that the two photons consisting of Planck particles bind to form the polarized electron.

Based on vacuum flow considerations, it was found that in the electron, the ratio of the number of rotation frequency of the Planck particles is the square of the number of rotations of the electron in the Compton in the same time. The distance the photons can travel around the perimeter is then, $v_{PL} = n_e^2$ cycles / sec, and the velocity is

$$\rightarrow \quad v_{PL} = \sqrt{2}\lambda_{PL} \times n_e^2 \quad cm / sec.$$

The ratio of the distance traveled for the photon rotation to the orbital Compton rotation is:

$$\sqrt{2}\lambda_{PL} n_e^2 / \lambda_e n_e = \mathfrak{R}_0 / \lambda_e$$

Where n_e is the number of revolutions the Planck particle makes in the time it takes for a single orbit around the Compton radius, and $\mathfrak{R}_0 = \sqrt{2}\lambda_{PL} n_e$ is the distance traveled. $\mathfrak{R}_0$ is thus the most probable orbiting radius, and $\lambda_e = \hbar / m_e c_0$ is the radius at which the electrons have an action equal to $\hbar$.

Content

Preface

The flow functions ϕ are vector creations of conjugate photons and, for this presentation, will be identified as up and down vectors as $\phi* \downarrow$ and $\phi \uparrow$, and identified as conjugate flow functions. These density flows are more mathematically proper in the Geometric Algebra, as described in [10], but are not as descriptive.

Interaction Kernels $\phi*\phi$

The interaction of the electron flow density in Eq.(2), with its conjugate, and shown in Fig. 3, is the probability density of two opposite Planck action flows that create the probability density of particle-particle intersection, resulting in a decrease in the action flow between the $\pm$ kernels.

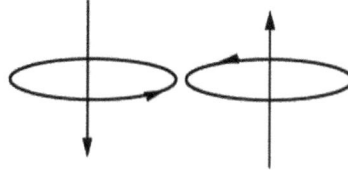

Fig. 2

The probability density of particle-particle intersection of the particles flow is the product of the opposing flow density streams on the most probable path, and for two bound $\pm$ electrons found and discussed in earlier works in Eq.(2), is:

$$\frac{\Delta\varepsilon}{m_e} = \left(\frac{\mathfrak{R}_0}{\lambda_e}\frac{1}{g_A^2}\right)\left(\left(\frac{\mathfrak{R}_0}{\lambda_e}\frac{1}{g_A^2}\right)\right) = \left(\frac{\alpha}{n_R}\right)^2 \rightarrow (\phi*\phi) \qquad (2)$$

These scalar relations were developed earlier before the flow vectors and integrals were understood.

4

The product of $(\phi * \phi)$ is the scalar energy created when the flow vector functions bind, and the energy is the ratio of the binding energy to the state energy of the free electron.

$$(\phi * \phi) = \frac{\text{State Energy Created}}{\text{Free Electron State Energy } m_e c_0^2} \quad (3)$$

Eq. (2), exactly defines the energy of the Rydberg states without reference to energy fields or charge. The interaction or product of the conjugate flow vectors at Compton radius creates an energy level exactly equal to the ratio of the fine structure constant to the Rydberg integer. This is the energy ratio of the state energy to the energy of the electron, and the state values defined are the Rydberg atomic states.

The energy of a state composed of the product of multiple conjugate flow functions will be shown to be:

$$\phi_x * \phi_x = (\phi_1 * \phi_1)(\phi_2 * \phi_2) \cdots \rightarrow \left(\frac{\varepsilon_x}{m_e}\right) = \sqrt{\left(\frac{m_1}{m_e}\right)^2 \left(\frac{m_2}{m_e}\right)^2 \left(\frac{m_3}{m_e}\right)^2 \cdots} \quad (4)$$

Feynman Action Contour Integrals and Vertex Kernels

The flow densities shown above can be readily identified as the Feynman contour flow densities of the electron. The linear and contour Feynman integrals over all paths are illustrated in Fig. 3 and 4.

Contour loops of bound particles   Linear free particles

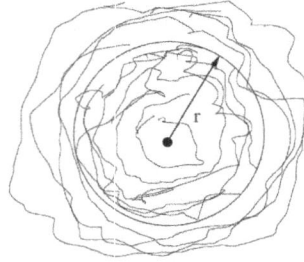 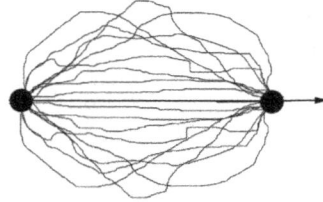

Fig. 3   **Fig. 4**

Feynman Action Flow Integrals

The Feynman integral action for rotating bound photons is similar to the linear flow integral, except it repeats, giving a periodic a space-time vertex kernel.

The momentum vector of the electron consists of the sum of the four derivative of the wavefunctions of two bound conjugate photons. Using the Dirac matrix of Geometrical Algebra, (GA) form of the electron photons shown earlier in [10], and they are:

$$\frac{\boldsymbol{p_1}}{\hbar} = \left[\gamma^\mu \frac{\partial}{\partial x^\mu} e^{\,i\left(\frac{m_1 c_0}{\hbar}\left(k \bullet x - c_0 t\right)\right)} \right] \qquad \frac{\boldsymbol{p_2}}{\hbar} = \left[\gamma^\mu \frac{\partial}{\partial x^\mu} e^{\,i\left(\frac{m_2 c_0}{\hbar}\left(-k \bullet x - c_0 t\right)\right)} \right] \quad (5)$$

6

The sum is:

$$p_1 + p_2 = \left\| \left(\gamma^\mu \left(\frac{m_1 c_0}{\hbar} - \frac{m_1 c_0}{\hbar} \right) - \gamma^k \frac{m_1 \Delta c_1}{\hbar} \right) \right\| = \left(\frac{m_e \Delta c_1}{\hbar} \right) \quad (6)$$

The magnitude of two four vectors has two polarizations that can be represented as up and down vectors:

$$\frac{p_1 + p_2}{\hbar} \rightarrow \frac{m_e c_0}{\hbar} \downarrow \text{ or } \frac{m_e c_0}{\hbar} \uparrow \quad (7)$$

The flow vectors found in Eq.(2), by vacuum flow considerations, have been identified as the Feynman contour action flow of the two photons in the electron on the most probable path of the rotating classical electron orbit $\mathfrak{R}_0$.

The action flow can be presumed to be photon momentum vectors, not a summation of imaginary phases, but a probability density of photon flow.

The Feynman contour flow integral vector around the most probable radius $\mathfrak{R}_0$, noted in Eq.(1), is:

$$\phi* = \frac{\displaystyle\int_0^{\mathfrak{R}_0/n_R} m_e c_0 dr}{\hbar} = \frac{\mathfrak{R}_0 m_e c_0}{\hbar g_A^2} \downarrow = \frac{S_e / n_R}{\hbar g_A^2} \uparrow = \frac{\alpha}{n_R} \quad (8)$$

The Rydberg integer, n_R, reduces the action of the free electron state, S_e, from the free particle value.

The product of this flow vector and its conjugate up vector is the state scalar energy Eq.(3)

$$\phi*\phi = \frac{S_e / n_R}{\hbar g_A^2} \downarrow \times \frac{S_e / n_R}{\hbar g_A^2} \uparrow = \left(\frac{\alpha}{n_R} \right)^2 = \frac{\Delta\varepsilon}{m_e c_0^2} \quad (9)$$

The integer n_R is the Rydberg integer and multiplier of the action quantum, and g_A^2 is the QED Anomalous Gyromagnetic factor, of the rotating photons that is present when all possible action paths are included. (See Appendix III.)

Eq. (9), defines the atomic Rydberg energy states without reference to charge or energy fields.

Kernels

Atomic Kernels

Pictorially, the created energy at the Compton radius can be viewed as the action decrease of the free electron, as the result of the integral around an increasing radius $\lambda_e n_R$.

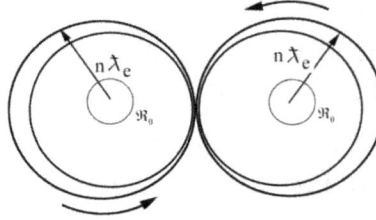

Fig. 5

Fig. 5 shows the binding of the two vertex kernels in Eq. (8), at the most probable path. This view is based on the expanding Compton radius of the energy created by the interaction in the atomic states.

The product of the flow densities of the two conjugate atomic kernels electrons bound at the Electron Compton radius is then:

$$\phi_A * \phi_A = \left(\frac{S_e / n_R}{\hbar g_A^2} \right)^* \left(\frac{S_e / n_R}{\hbar g_A^2} \right) = \left(\frac{\alpha}{n_R} \right)^2 = \frac{\Delta \varepsilon}{m_e} = K_A * K_A \qquad (10)$$

This product of the two conjugate flow functions is the scalar binding energy of the specific Rydberg states and will be identified as Atomic Kernels.

There is another group of action radii for the bound electron that are integral values of the action quantum. Which are reciprocal of the Rydberg levels or ratios of action quantum $\hbar$ to the action S_e. These are the action flow of electrons in rotation radii inside the electron radius.

These states will be referred to as nuclear states, and the kernels as Nuclear Kernels. The ratio of the action quantum with the particle action is:

$$\phi_N = \frac{\hbar}{\mathfrak{R}_0/n_R \int\limits_0^{} m_e c_0 dr} = \frac{\hbar\eta}{\left(S_e/n_R\right)} = \frac{\varepsilon}{m_e} = \frac{n_R\eta}{\alpha} \quad (11)$$

Atomic states action ratios are thought of as multiples of $n_R\hbar$, as an increase in the Compton radius (Eq. (2)). The interior action ratios are the reciprocal of the same ratio, $n\hbar$. Instead of the action being multiple values of the action quantum, the states appear as fractions of a single value of the action quantum. The all-path integral includes the nuclear QED anomalous ratios η.

Pictorially, the created mass can be viewed as action flow density as ratio electron radius to fractions of that radius, as shown in Fig. 6. In both cases, interior and exterior, the actions S are integral values of $\hbar$.

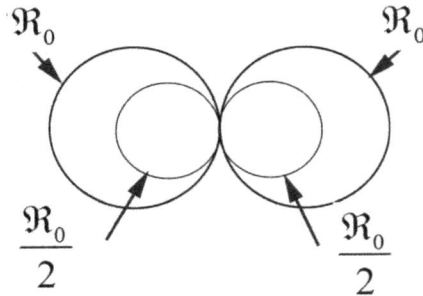

Fig. 6

The corresponding expression to Eq. (10), of the exterior binding, for the interior states is the Nuclear Vertex function:

$$\phi_N {}^*\phi_N = \left(\frac{\hbar\,\eta}{(S_e/n_R)}\right)^* \left(\frac{\hbar\,\eta}{(S_e/n_R)}\right) = \frac{\Delta\varepsilon}{m_e} = K_N {}^*K_N \qquad (12)$$

This is the energy required by the binding of the two nuclear kernels that could be referred to as quarks. As for the atomic states, this is the energy radiated away when it is created and thus must be put in when the particle is formed. The energy expressed as mass in Eq. (12) is a deficit energy. (See Appendix VI.)

The consilience between Eq. (10), and Eq. (12) is compelling, if not decisive, in theoretical merit to the flow probability density concept.

Mechanics, Binding, and the Vertex Kernel

Eq. (10) and Eq. (12) are respectively the binding energy of a pair of atomic and nuclear vertex. Including the QED effect from Eq. (28), and Eq. (27), these have been identified as atomic and nuclear vertex kernels.

$$\phi_A {}^*\phi_A = \left(\frac{S_e/n_R}{\hbar\,g_A^2}\right)^* \left(\frac{S_e/n_R}{\hbar\,g_A^2}\right) \rightarrow \frac{\Delta\varepsilon}{m_e} = K_A {}^*K_A$$

$$\phi_N {}^*\phi_N = \left(\frac{\hbar\,\eta}{(S_e/n_R)}\right)^* \left(\frac{\hbar\,\eta}{(S_e/n_R)}\right) \rightarrow \frac{\Delta\varepsilon}{m_e} = K_N {}^*K_N$$

$$(13)$$

Nuclear Multiple Kernel Vertex Function

The atomic vertex represents the Rydberg states and, as shown in Fig. 4, are generally only two vertex kernels in a vertex function. The nuclear kernels can bind inside the nucleus in any number to form complex particles, as will be illustrated in the paper on vertex functions. When bound together, this becomes a vertex function of the Feynman path integrals.

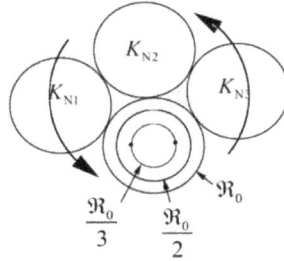

Fig.7

The nuclear kernels can bind in any number of multiple pairs since pairs of up and down bound kernels have no polarization. Any group of vertex kernels can be factored into two products, one with a positive kernel and one with a negative kernel, thus implying the effect of a positive and negative charge as shown in Eq. (14).

$$K_P^* K_P = \left(K_1^* K_1\right)\left(K_2^* K_2\right)\left(K_3^* K_3\right)\bullet\bullet \qquad \text{a.}$$

$$K_P^* = K_1^* \sqrt{\left(K_2^* K_2\right)\left(K_3^* K_3\right)\bullet\bullet\bullet} \qquad \text{b.} \qquad (14)$$

$$K_{P1} = K_1 \sqrt{\left(K_2^* K_2\right)\left(K_3^* K_3\right)\bullet\bullet\bullet} \qquad \text{c.}$$

The positive and negative kernels are positive and negative vector flow particles.

Eq. (14) a. shows a neutral vertex function as a group of un-aligned kernel pairs, and Eq. (14), b. and c. show positive and negative flow vector particles having a single polarized particle with a group of unaligned pairs. (a. defines a is a neutral particle, and b. & c. define positive and negative charged particles.)

Binding and Factoring of Nuclear Vertex Functions

In QM, the superposition or sum of QM wavefunctions ψ creates the complete function from an infinite sum of sub-functions. The summation or superposition's of states creates the final state, or energy level.

Vertex kernels are the product of an infinite number of factors that can be divided into a series of smaller factors that make the total action and mass. Each factor can be considered a mass and thus the root of a vertex can be considered the product of two masses.

Vertex kernels can be factored into an infinite number of sub-particles so long as the action is an integral value of the action quantum.

$$K_A = \frac{m_1}{m_e} \frac{m_2}{m_e} \frac{m_3}{m_e} \frac{m_4}{m_e} \frac{m_5}{m_e} \cdots$$

$$K_A = \frac{m}{m_e} = \sqrt{\frac{m}{m_e}} \times \sqrt{\frac{m}{m_e}} = \sqrt{\sqrt{\frac{m}{m_e}}} \times \sqrt{\sqrt{\frac{m}{m_e}}} \times \sqrt{\sqrt{\frac{m}{m_e}}} \times \sqrt{\sqrt{\frac{m}{m_e}}}$$

(15)

Irrational Numbers

A kernel that is part of a bound mass has to have an integral value of $n \hbar$; thus, a kernel that has a value $\left(\dfrac{S_e}{\sqrt{2}\hbar} \right)$ could not bind, but two particles such that $\rightarrow \sqrt{\left(\dfrac{S_e}{2\hbar} \right)^*} \sqrt{\left(\dfrac{S_e}{1\hbar} \right)}$ has the same mass, and both have $n \hbar$ integral action and can bind as a neutral unaligned pair in a vertex function.

Energy and Mass Binding Examples

Positronium Atomic

Noting the center of mass orbiting radius of the $\pm$ electrons in positronium is half the Compton radius of the electron, the action of Eq. (10) is reduced by half, but the integer of $\hbar$ is not changed. This is the proper interaction energy of a positive and negative electron establishing the Rydberg energy levels without the concept of charge, and is the exact energy levels of positronium, 6.803ev.

$$\left(K_A * K_A \right) = \left(\frac{S_e / 2}{n_R \, \hbar \, g_A^2} \right)^* \left(\frac{S_e / 2}{n_R \, \hbar \, g_A^2} \right) = \frac{1}{4} \left(\frac{\alpha}{n_R} \right)^2 = \frac{\Delta \varepsilon}{m_e} \rightarrow \quad = 6.803 \text{ev} / n_R$$

$$(16)$$

Muon Nuclear

The muon vertex function identified in [25] was found to be the product of four nuclear kernels, three of which have the same value of $n_R = 16$:

$$\left(\frac{m_\mu}{m_e}^* \frac{m_\mu}{m_e} \right) = \left[\left(\frac{\hbar}{S} \frac{\eta}{2^4} \right)^* \left(\frac{\hbar}{S} \frac{\eta}{2^4} \right) \right] \left[\left(\frac{\hbar}{S} \frac{\eta}{2^3} \right)^* \left(\frac{\hbar}{S} \frac{\eta}{2^4} \right) \right] \qquad (17)$$

This is two pairs of two bound photons [25], forming a positive and negative muon pair.

Each of the Muon conjugate vertex pairs is neutral and can be factored into positive and negative Muon vector flow functions. Each Muon has a single odd vertex with a bound pair being an unaligned neutral vertex, as shown in Eq. (18):

$$\phi_\mu = + \frac{m_\mu}{m_e} = \left(\frac{\hbar}{S_e} \frac{\eta}{2^4} \right) \left(\sqrt{\left(\frac{\hbar}{S_e} \frac{\eta}{2^3} \right)^*} \sqrt{\left(\frac{\hbar}{S_e} \frac{\eta}{2^4} \right)} \right) = 206.7674344$$

$$\phi_\mu^* = - \frac{m_\mu}{m_e} = \left(\frac{\hbar}{S_e} \frac{\eta}{2^4} \right)^* \left(\sqrt{\left(\frac{\hbar}{S_e} \frac{\eta}{2^3} \right)^*} \sqrt{\left(\frac{\hbar}{S_e} \frac{\eta}{2^4} \right)} \right) = 206.7674344$$

$$(18)$$

Several other mass values of particles and vertex functions are illustrated in the paper on vertex functions. [25]

Gravitation—Non-Aligned Particle Densities

The mystery of the gravitation connection to electricity has vexed theorists since Euler developed differential field theory. The energy required by the fictitious fields of gravitation is beyond calculable. Δc Mechanics for gravitation, however, is much simpler than that for electron interactions because there are no states. The probability density generated by non-aligned particles is totally random, and the only forces are those associated with gradients in the probability density.

The Difference between Electric and Gravitational Interaction

The electron and the kernel of all particles are composed of rotating photons. Its rotation probability density lies in a plane, and the generated density in that plane repeats at ν_e.

The repeating exterior probability flow is still present, but the rotation plane conjugate pairs are randomly oriented and at a point in space such that it is not repeating at the electron frequency ν_e. At any point, the orientation is random and there is an omnidirectional increase in the vacuum density and a decrease in the velocity of light. This reduces the effect of the interaction by ν_e^2, and since the electron rotation frequency is 1.325E-20 /sec, the reduction effect is about 1/1.75695025e40. This constitutes the difference in the magnitude of electricity and gravitation. The value of gravitational interaction can be demonstrated by the probable flow densities.

Gravitational Flow Density Functions

The flow probability density of a kernel at the Compton radius is:

$$P = \frac{\sqrt{2}\lambda_{PL} v_e}{\lambda_e} \tag{19}$$

A non-aligned kernel or pair of kernels not rotating in a plane and not aligned is free to be at any angle:

$$P_1 = \frac{\sqrt{2}\lambda_{PL} v_e}{\lambda_e} \rightarrow \frac{\sqrt{2}\lambda_{PL}}{\lambda_e} \tag{20}$$

If the distance between the radius of the mass and the other particle is not at the Compton radius but at a distance r, then the density is:

$$\phi_r = \frac{\sqrt{2}\lambda_{PL}}{r} \qquad @\, r = \sqrt{2}\lambda_{PL} \rightarrow 1 \tag{21}$$

This is not a function of the mass of a second particle but the value of the probability density of the first particle as a function of r.

The product of the two functions is then:

$$\phi_m \phi_r = P_1 \frac{\sqrt{2}\lambda_{PL}}{r} \qquad @\, r = \sqrt{2}\lambda_{PL} \rightarrow \phi_m \phi_r = P_1 \tag{22}$$

Inserting the specifics gives:

$$\phi_m \phi_r = \left(\frac{\sqrt{2}\lambda_{PL} m_m c_0}{\lambda_m}\right) \times \left(\frac{\sqrt{2}\lambda_{PL}}{r}\right) = \frac{S_{PL}}{\hbar}\frac{\sqrt{2}\lambda_{PL}}{r} = \frac{2Gm_m}{c^2 r} = \frac{\Delta c}{c_0} \tag{23}$$

The Planck radius is $\lambda_{PL} = \sqrt{G\hbar/c^3}$

The change in the velocity of light at the position r from the particle is the ratio of the gravitational radius to the distance, or the gravitational potential:

$$\phi_m \phi_r = \frac{2Gm_m}{c^2 r} = \frac{\Delta c}{c_0} = \frac{\Delta \varepsilon}{\varepsilon_0} \tag{24}$$

This is the omnidirectional flow probability density at any point at a distance r from the mass particle, thus increasing the ubiquitous omnidirectional vacuum flow density and lowering the velocity of light:

$$\frac{\Delta c}{c_0} = \frac{c_0 - c}{c_0} = \left(\frac{2\mu_m}{r}\right) \quad \rightarrow c = c_0\left(1 - \frac{2\mu_m}{r}\right) \tag{25}$$

This relation is the Roger Blandford, Kip Thorn, velocity of light from a mass in GR as projected onto flat space [15], and can be summed over any number of particles. The results thus match the physical results of GR for the x particle located at r.

From Eq. (25), the ratio of the potential energy to the total energy is then:

$$\frac{m_x c_0^2 - m_x c c_0}{m_x c_0^2} = \left(\frac{2\mu_m}{r}\right) \quad \rightarrow \Delta\varepsilon = \left(m_x c_0^2 - \frac{2Gm_m m_x}{r}\right) \tag{26}$$

Thus, the same change in energy of a mass particle, experiencing probability flow density of another mass, is the same as the change in the gravitational potential.

Δc Mechanics thus defines the mechanism of gravitation as well as electrical interactions.

Conclusion

The flow vector density functions are a new approach to quantum mechanics without the baggage of fictitious energy fields and inexplicable charge.

The Feynman flow probability functions are conceptually understandable and provide state interactions, replacing the need for charge and field constructs that are fictitious and mathematically problematic. They provide concrete values of particle mass without inserting arbitrary coupling constants.

Whether the presentation here is as well done as it should be is open to discussion; however, the process is undeniable.

This is an alternate basis for quantum mechanics, which is a purely mathematical structure, not directly related to anything observable, and the amplitude is not definable. QM makes up the missing interaction by adding an ad hoc charge and field energy that has no tangible basis.

The product of the vertex kernels defines the energy of the atomic states as well as the mass of the nuclear states to a very high precision and, as such, this theory is very much in concert with Ellington's conjecture that particle mass should be defined by the product of masses rather than the sum [13].

> "It was an intuition without a precise justification," said Adán Cabello, a quantum theorist at the University of Seville in Spain. "But it worked." and yet for the past 90 years and more, no one has been able to explain why. [16]

Author's Note

The development of this theory has been a lengthy process, and it is far from done. The content and the divergence of this development from standard QM is revolutionary, and the calculated numerical values of mass particles are the proof of concept.

It was only recently that the vertex vector flow functions were identified, but finally realizing the connection of the flow density probability with the Feynman path integrals brought the concept of Δc Mechanics to a better theoretical foundation. Many of the math nuances are not finished, but the basis is defined. The theory does have revolutionary aspects, but at this time there is not much interest, someday someone will pick it up and move it along. This author is not likely to be around for that.

References

The results of the development of Δc theory can be found in a number of papers that are perhaps not as clear in the definitions presented here, primarily because the author did not understand as well at the time it was written. References 20–25 are newer and perhaps have better clarity. Most of the references in current papers are to the authors papers, which contain the primary references to the work.

A. DT Froedge, *The Physics of Delta-c Mechanics*, ISBN-13: 979-8218347178 (Feb. 14, 2024) (Papers referenced can be found as chapters in this publication, and available online in DOI numbers).

1. DT Froedge, "The Concepts and Principles of Delta-c Mechanics," August 2024, DOI: 10.13140/RG.2.2.31630.37445, https://www.researchgate.net/publication/382850589.

2. DT Froedge, "The Connection between Electric Charge, Gravitation, and the Feynman Sum over All Histories View of Quantum Electrodynamics," April 2020 Conference: APS April 18-21, 2020, Washington, DC, https://absuploads.aps.org/presentation.cfm?pid=18355, https://www.researchgate.net/publication/341310206.

3. DT Froedge, "The Electron as a Composition of Two Vacuum Polarization Confined Photons," April 2021, DOI: 10.13140/RG.2.2.18971.18722, https://www.researchgate.net/publication/350740864.

4. DT Froedge, "The Fine Structure Constant from the Feynman Path Integrals," March 2021, DOI:10.13140/RG.2.2.12979.55846, https://www.researchgate.net/publication/350188862.

5. DT Froedge, "Vacuum Polarization, Gravitation, Charge, and the Speed of Light," Sept. 2021, DOI:10.13140/RG.2.2.15619.22569. https://www.researchgate.net/publication/354474157 (Equations 20–33).

6. DT Froedge, "The Calculated value of the Fine Structure Constant from Fundamental Constants," September 2021, DOI:10.13140/ RG.2.2.34349.41440.

7. "CODATA values of the fundamental physical constants," https:// www.nist.gov/programs-projects/codata-values-fundamental-physical-constants.

8. DT Froedge, "Structure of Elementary Nuclear Particles in Delta-c Mechanics," https://www.researchgate.net/publication/385782943, November 2023, DOI: 10.13140/RG.2.2.27181.70884.

9. DT Froedge, "Vacuum Polarization, Gravitation, Charge, and the Speed of Light," Sept. 2021, DOI:10.13140/RG.2.2.15619.22569, https://www.researchgate.net/publication/354474157.

10. DT Froedge, "The Dirac Equation and the Two Photon Model of the Electron revised," April 2021, DOI: 10.13140/RG.2.2.19095.70564, https://www.researchgate.net/publication/350922403.

12. DT Froedge, "The Dirac Equation and the two Photon Model of the Electron revised," April 2021, DOI: 10.13140/RG.2.2.19095.70564 https://www.researchgate.net/publication/350922403.

13. Arthur Eddington, *Relativity Theory of Protons and Electrons*, New York: The MacMillan Company Cambridge England, The University Press, 1936

14. He, et al., "Optomechanical measurement of photon spin angular momentum and optical torque in integrated photonic devices," Sept. 9, 2016, *Science Advances* Vol. 2, no. 9, DOI: 10.1126/sciadv.1600485.

15. Roger Blandford and Kip S. Thorne, "Applications of Classical Physics," (in preparation, 2004), Chapter 25, http://pmaweb.caltech.edu/Courses/ph136/yr2012/1227.1.K.pdf.

16. The Quanta Newsletter, Feb. 13, 2019, https://www.quantamagazine.org/the-born-rule-has-been-derived-from-simple-physical-principles-20190213.

17. Schwinger, J. (1949), "II. Vacuum polarization and self-energy," Physical Review, Quantum Electrodynamics, 75 (4): 651–679.

18. DT Froedge, "Structure of Elementary Nuclear Particles in Delta-c Mechanics," November 2024, DOI:10.13140/RG.2.2.27181.70884, https://www.researchgate.net/publication/385782943.

Recent Papers

The following recent papers by the author focus on the application, mechanism, and relation to current theory as well as physical applications:

1a. DT Froedge, "Calculated Atomic Masses in Delta-c Mechanics," January 2025, DOI: 10.13140/RG.2.2.30292.51842, https://www.researchgate.net/publication/388219158.

2a. DT Froedge, "Structure of Elementary Nuclear Particles in Delta-c Mechanics," November 2024, DOI: 10.13140/RG.2.2.27181.70884, https://www.researchgate.net/publication/385782943.

22. DT Froedge, "The Concepts and Principles of Delta-c Mechanics," August 2024, DOI: 10.13140/RG.2.2.31630.37445, www.researchgate.net/publication/382850589.

3a. DT Froedge, "Acceleration Gravitation and Origin of Centrifugal Force in Delta-c Mechanics," July 2024, DOI: 10.13140/RG.2.2.28944.42247, https://www.researchgate.net/publication/381915343.

4a. DT Froedge, "Relativistic Time Dilation Illustrated in Delta-c Mechanics, The Twin Paradox is an Illusion," December 2023, DOI: 10.13140/RG.2.2.15584.46085, www.researchgate.net/publication/376831390.

5a. DT Froedge, Feynman Vertex Functions, Researchgate, June 2025, DOI: 10.13140/RG.2.2.35247.24487, https://www.researchgate.net/publication/392331016

Appendix I

Conjugate Kernels

The factors in Eq. (10) and Eq. (12) can be referred to as conjugate kernels, and are each the probability density of two oppositely polarized photons. The asterisk does not represent an imaginary number, but one is a positive vertex kernel and the other a negative vertex kernel, with an opposite polarization. Conjugate ± particles, such as electron-positrons can bind due to the counter flow of the probability flow density. The polarization doesn't affect the value of the product but has internal effects on the kernels. Opposite polarized kernels bind in pairs, and become non-polarized, and alike pairs do not.

The conjugate vectors concept is presented here as a visual for understanding of the process. A more proper mathematic representation is presented in the Dirac matrix of Geometric Algebra in: "The Dirac Equation and the two Photon Model." [12]

$\phi * \phi$ $\phi * \phi *$

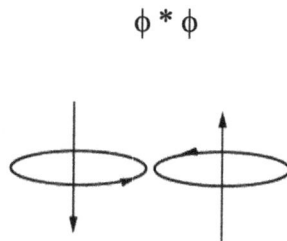 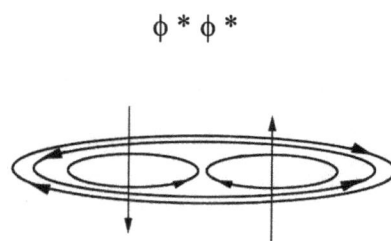

Fig. 2A   Fig. 2B

The Relationship with Charge

Fig. 2A shows a positive and negative kernel, both rotating clockwise with opposite B vectors. Between the two aligned particles is a volume of space where the action flow probability density decreases the velocity of light or increases the index of refraction. This increase in the index of refraction pulls the action paths from one particle to the other as an opposite charge. Fig. 2B shows a pair of identical particles with B vectors anti-aligned. The photon action flow in between the particles is in the same direction and does not interfere. On the exterior, however, there is a flow density interaction which pulls the particles apart, just as a repulsive charge would.

For illustration, opposite particles are marked with an asterisk, but the factors are not imaginary numbers but designating alike and opposite polarization or positive and negative vertex.

Appendix II

The Rydberg Integer $n_R \hbar$

The binding of vertex kernels all requires integral values $n \hbar$. This results from the fact that the rotating frequency of the kernels is proportional to $\hbar$. If the numbers are integers, then after a period of time the cycles repeat, whereas if they were irrational that could not be true.

Particles bind at the most probable action radius, and the relative rotation of each bound particle at the contact point must be an integral number of $n \hbar$, otherwise the binding cannot be not stable.

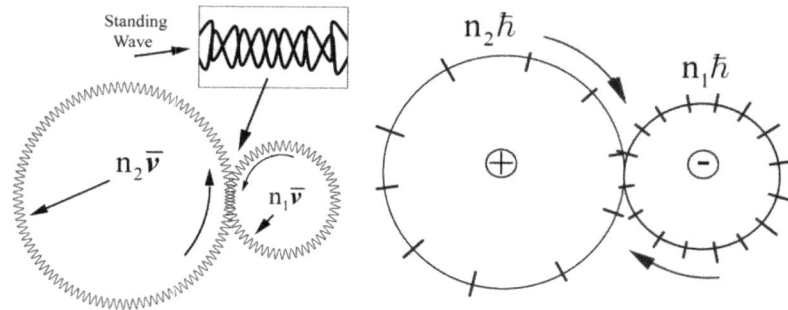

Fig. 6

The interaction of a particle having action as $n_1 \hbar$ with a particle having action $n_2 \hbar$ results in a standing wave of probability flow at the contact point that satisfies flow continuity relations and repeats as the particles counter-rotate. fi Standing waves between Kernels is the exchange of flow probability density that satisfies the continuity condition.

The integral value $n \hbar$ is thus a requirement for vertex kernels of bound particles.

Appendix III

Anomalous QED Ratios

There are two anomalous ratios that play a role in nuclear-atomic systems, shown in Eq. (10) and Eq. (12), and those are the factors g_A and η. They are the QED-determined loop integral delay for particles moving in bound orbits. The ratio has no effect on the free electron, but when an electron is bound in rotation in the Feynman action, integral loops cause a delay that alters the particle rotation.

The first ratios that apply to atomic or Rydberg levels is the Anomalous Gyromagnetic ratio g_A. When an electron is moving in free space, there are no loop delays; however, when it is in rotation in the Compton orbit the QED loop delays increase the radius, due to the QED time delay, and lower the binding energy. Its value has been theoretically predicted and measured to about twelve significant digits. The value is:

$$g_A = 1.0011596521807 \tag{27}$$

This factor is an offset of the Rydberg atomic levels.

The second is designated as Eta. η. It represents the QED effect on the rotating photons inside a nuclear vertex function.

$$\eta = 1.00060014721177 \tag{28}$$

The value of η has not yet been determined by QED methods, but it should not be dismissed as arbitrary, as it is small but crucial to the precise mass of several unconnected known particles. [9] If there is a fudge factor in this development, this is it. This ratio, like g_A, is not calculated from physical constants.

Appendix IV

Energy Deficit

The conjugate bound pairs in the nucleus are opposite and, if in free space, would annihilate. In binding, however, the electrons, like their atomic counterparts, have given up energy.

And, like the atomic counterparts, also have insufficient energy to escape. For a hydrogen atom, the state energy has to be kinetically injected to remove the electron. For a nuclear mass, the same principle applies; the energy of the entire mass has to be injected to remove the particle from the nucleus. This implies that nuclear mass exists as a deficit of energy and particles in the nucleus cannot annihilate unless energetically removed.

This gives another perspective to the Einstein energy relation, in that nuclear inertial mass exists as a deficit of energy.

$$m = \varepsilon/c^2 \qquad \rightarrow \qquad m = -\varepsilon/c^2 \qquad (29)$$

Feynman Vertex Functions

D.T. Froedge

v060225

Introduction

Vertex Functions are states of mass particles located at the end point, or final states of the Feynman linear action integrals, and represent the rest mass of a collection of Vertex Kernels. The Vertex Kernels defined here are the structure of the probability flow density of electrons in quantum states that bind together as multi-particle structures. The detail of the structure is developed in "Feynman Flow Density Alternative to Wavefunctions." [20]

The vertex kernels are solutions to the Feynman flow interaction, $\phi^*\phi$ of particle states that are alternates to the QM probability density state functions $\psi^*\psi$.

Both the Feynman path integrals and the Schrödinger equation reproduce the initial and final states of particle motion at the vertex of the Feynman diagrams. It is found that these are the vertex function and shown to be the binding energy of atomic states that define the rest mass of nuclear particles.

The vertex function of atomic states will be familiar, as are all states of the electron, but the nuclear states, which are also states of the electron, may not be as familiar. The vertex "function" is a product of vertex kernels that form the state values. It is illustrated that each quark of the proton is an electron in action states of $n\hbar$, and the product of the three quarks is the total mass.

The mass of the vertex function is the product of the vertex kernels and is consistent with Eddington's view of the mass of particles—like wavefunctions should be products. [13]

The energy of a state composed of the product of multiple conjugate flow functions will be shown to be:

$$\phi_x{}^*\phi_x = (\phi_1{}^*\phi_1)(\phi_2{}^*\phi_2)\bullet\bullet \;\rightarrow\; \left(\frac{\varepsilon_x}{m_e}\right) = \sqrt{\left(\frac{m_1}{m_e}\right)^2\left(\frac{m_2}{m_e}\right)^2\left(\frac{m_3}{m_e}\right)^2\bullet\bullet} \quad (1)$$

The Feynman vector flow functions ϕ_x and $\phi_x{}^*$ for particular states of action will be referred to as Vertex Kernels $(K_1){}^*(K_2)$, and the product as vertex functions.

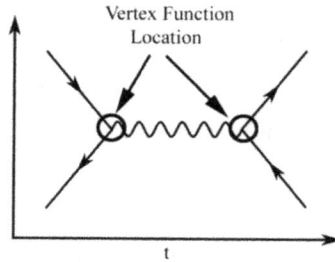

Fig. 1: Vertex Positions

Flow Probability Density

The Vertex Kernel is the vector flow density of a state of a rotating electron. As the two photons rotate inside the electron, there is a probability via the Feynman postulates of the circularly flow location existing throughout space. This gives the exterior probability flow density in a plane on the action paths as a function of the distance from the center of momentum. The product of the Kernels is the mass energy reference to the mass energy of the free electron.

$$\left[(K_1)*(K_2)\bullet\bullet\bullet\right] \rightarrow m_1^2 \; m_2^2 \; m_3^2 \bullet\bullet\bullet \tag{2}$$

Conjugate Particles

Kernels bind in pairs, as shown in Fig. 2, and can be referred to as conjugate kernels. Each is the probability density of two oppositely polarized kernels. The asterisk does not represent a negative number, but one is an electron and the other a negative electron (positron), with an opposite polarization and flow density, as shown in Fig. 2A. [3]

The conjugate vectors concept is presented here as a visual for understanding of the process. A more proper mathematic representation is presented in the Dirac matrix of Geometric Algebra in: "The Dirac Equation and the two Photon Model." [12]

Conjugate particles are opposite ± particles, such as electron-positron that bind due to the counter flow of the probability density:

$$\phi * \phi \qquad\qquad \phi * \phi *$$

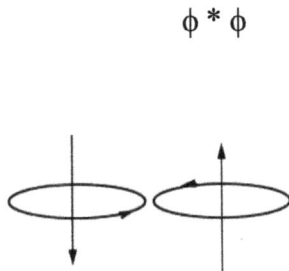

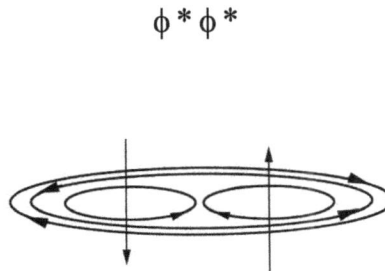

Fig. 2A   **Fig. 2B**

Particle-Particle Descriptive

The photons in an electron rotate such that the electric vectors are constant along the radial vector, and the magnetic vector is perpendicular to the rotation plane.

Fig. 2A shows a positive and negative particle, both rotating clockwise with opposite B vectors. Between the two aligned particles is a volume of space where the action flow probability density decreases the velocity of light or increases the index of refraction. This increase in the index of refraction pulls the action paths from one particle to the other, simulating a charge effect. Fig. 2B shows a pair of identical particles with B vectors anti-aligned. The photon action flow in between the particles is in the same direction and does not interfere. On the exterior, however, there is a flow density interaction that pulls the particles apart, just as would a repulsive charge.

The product of a conjugate pair cancels the up and down ($\uparrow\downarrow$) time-dependent vectors, yielding a scalar energy ratio. A bound pair is not aligned with other parties in a vertex function or an external field. A group of vertex kernels can only have one aligned kernel, and a neutral vertex function will have no unaligned pairs. Eq. (3) illustrates a neutral pair and the factoring, showing the positive Vertex function.

$$\phi_x * \phi_x = (\phi_1 * \phi_1)(\phi_2 * \phi_2)_{\bullet\bullet} \rightarrow +\phi_x^* = +\phi_1^* \sqrt{\left(\phi_2^*\phi_1\right)^2 \left(\phi_3^*\phi_3\right)^2}_{\bullet\bullet}$$

$$(3)$$

Vertex kernels are higher action states of free electrons, and the interaction energy and mass of the vertex kernels are all expressed as ratios of the mass of the free electron $m = m_x / m_e$.

Flow Functions of Atomic and Nuclear Kernels

From earlier work, it was found that the ratio of the creation radius of rotating photons that form the electron is the same as the classical radius (See Appendix I). Eq.(4) shows the equivalent relations:

$$\phi_e = \frac{\sqrt{2}\lambda_{PL}\bar{v}_e}{\lambda_e} = \frac{\mathfrak{R}_0}{\lambda_e} = \frac{\mathfrak{R}_0 m_e c_0}{\hbar} = \frac{S_e}{\hbar} = \alpha \tag{4}$$

First is the relation with the Planck particle; second, the relation between the electron radius $\mathfrak{R}_0 = \sqrt{2}\lambda_{PL}\bar{v}_e$ and the electron Compton radius λ_e; third & fourth, the free electron action; and last, the fine structure constant. (See [26] for development.)

The internal action of the free electron is $S_e = \mathfrak{R}_0 m_e c_0$.

Vertex Kernels

Vertex Kernels are all composed of action states of the electron, and for vertex kernels there are two ranges of action states that are possible. One is for atomic states with the action ratio being the ratio of the free electron action to the action quantum $\hbar$ and having a value less than one:

$$\phi_A = \frac{S_e / n_R}{\hbar} < 1 \tag{5}$$

The second is for nuclear states with a density flow function that is the reciprocal of the atomic kernel. It is the ratio of the action quantum $\hbar$ to the free electron action and has a value greater than one:

$$\phi_N = \frac{\hbar}{S_e / n_R} > 1 \tag{6}$$

The flow functions of the atomic and nuclear states are reciprocals except for the QED effects. (See Appendix II.)

The two vertex kernels stated in Eq (5) and Eq (6) then are:

$$\text{(A)} \qquad \phi_A = \frac{\mathfrak{R}_0}{\lambda_e n_R} = \frac{(\mathfrak{R}_0 m_e c_0 / n_R)}{\hbar} = \frac{S_e / n_R}{\hbar}$$

$$\text{(B)} \qquad \phi_N = \frac{\lambda_e n_R}{\mathfrak{R}_0} = \frac{\hbar}{(\mathfrak{R}_0 m_e c_0 / n_R)} = \frac{\hbar}{S_e / n_R}$$

$$\tag{7}$$

This can be thought of as the ratio of the action inside and outside the electron Compton radius.

Feynman Flow Function

The functions of Eq.(7) were originally found by a more classical approach to particle interaction by flow density and photon-photon mean free paths. It has recently been realized that flow functions are just the Feynman contour integrals of the action paths of the probability flow of the rotating photons. The electron radius originally found in fundamental units $\mathfrak{R}_0$ is the classical electron radius $\mathfrak{R}_0 = \alpha \lambda_e g_A^2$ with the inclusion of the QED anomalous g_A factor.

The specific values for the flow functions ϕ in Eq.(7), located at the vertex of the Feynman diagrams, are referred to as vertex kernels. The binding of the atomic kernels establishes the well-known Rydberg energy states. The energy levels of the nuclear kernels are not well known but do establish the well-known mass of the nuclear particles.

Feynman Action Flow Integrals

The Feynman integral action for rotating bound photons is similar to the linear flow integral except it repeats, giving a periodic space-time vertex kernel.

The momentum vector of the electron consists of the sum of the four derivative of the wavefunctions of two bound conjugate photons. Using the Dirac matrix of Geometrical Algebra, (GA) form of the electron photons shown earlier in [10], and they are:

$$\frac{\boldsymbol{p_1}}{\hbar} = \left[\gamma^\mu \frac{\partial}{\partial x^\mu} e^{i\left(\frac{m_1 c_0}{\hbar}(k \bullet x - c_0 t)\right)} \right] \qquad \frac{\boldsymbol{p_2}}{\hbar} = \left[\gamma^\mu \frac{\partial}{\partial x^\mu} e^{i\left(\frac{m_2 c_0}{\hbar}(-k \bullet x - c_0 t)\right)} \right]$$

$$(8)$$

The magnitude of two four vectors have two polarizations that can be represented as up and down vectors:

$$\frac{\boldsymbol{p_1} + \boldsymbol{p_2}}{\hbar} \rightarrow \frac{m_e c_0}{\hbar} \downarrow \text{ or } \frac{m_e c_0}{\hbar} \uparrow \qquad (9)$$

Feynman Atomic Kernels

The action flow ratio for atomic particles is the Feynman contour integral of the action around the most probable path. From Eq. (7), this is: $(\mathfrak{R}_0 m_e c_0 / n_R)/\hbar$

The Feynman integral of action around the most probable path is then:

$$\phi_A = \frac{\displaystyle\int_0^{\mathfrak{R}_0/n_R} m_e c_0 dr}{\hbar} = \frac{\mathfrak{R}_0 m_e c_0}{n_R \hbar} \rightarrow \frac{S_e}{n_R \hbar g_A^2} = \frac{\alpha}{n_R} \qquad (10)$$

This is the Feynman vertex kernel for the Rydberg atomic states.

The QED anomalous spin factor g_A^2, noted in Eq. (27) , see Appendix II, has been included.

Pictorially, this can be viewed from the perspective of the Compton radius of the binding energy, showing the electron action divided by n times the action quantum $n_R \hbar$, shown in Fig. 3 below.

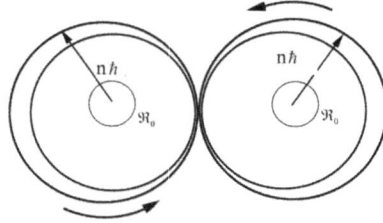 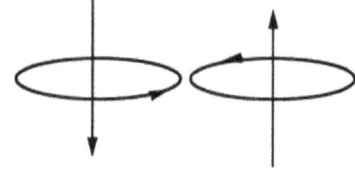

Fig. 3a   **Fig. 3b**

The product of the ± electron flow functions is then the Rydberg energy states:

$$\phi_A * \phi_A = \left(\frac{S_e}{n_R \hbar g_A^2}\right)^* \downarrow \left(\frac{S_e}{n_R \hbar g_A^2}\right)\uparrow = \frac{\Delta\varepsilon}{m_e} = K_A * K_A = \left(\frac{\alpha}{n_R}\right)^2$$

$$(11)$$

This is the product of two opposite flow function vectors and yields the well-known Rydberg energy states of the interaction.

The explanation of the integral values of the Rydberg integer n_R for the binding of kernels is discussed in Appendix III.

Nuclear Kernels

The action flow vector ratio for nuclear particles is the Feynman contour integral of the action around the most probable path. From Eq. (4), this is: $\Re_0 / n_R$.

The Feynman integral of action around the most probable path for a nuclear kernel is then:

$$\phi_N = \frac{\hbar}{\dfrac{\mathfrak{R}_0/n_R}{\displaystyle\int_0^{} m_e c_0 dr}} = \frac{\hbar}{S_e / n_R} \rightarrow \frac{\hbar\,\eta}{S_e / n_R} \qquad (?)$$

This is the Feynman vertex kernel for the nuclear states, including the QED spin factor η from the all space integral Eq.(28), (see Appendix II).

Atomic states are action ratios as multiples of $n\hbar$ (Eq. (11)). The nuclear action ratios are the reciprocal of the same ratio $n_R\,\hbar$.

Pictorially, this can be viewed from the perspective of the Compton radius of the binding energy showing the flow density function ϕ_N as ratio electron radius to fractions of $\mathfrak{R}_0 / n_R$, shown in Fig. 4 below.

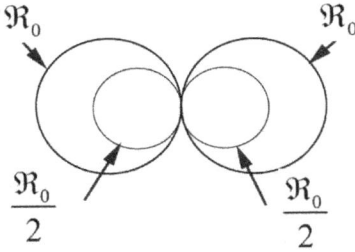

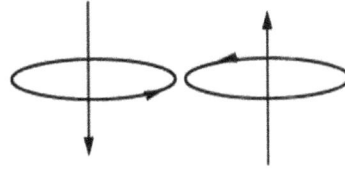

Fig. 4a   **Fig. 4b**

The corresponding expression to Eq. (11) for the nuclear kernels is:

$$\phi_N {}^* \phi_N = \left(\frac{\hbar\,\eta}{(S_e / n_R)}\right)^{\!*} \left(\frac{\hbar\,\eta}{(S_e / n_R)}\right) = \frac{\Delta\varepsilon}{m_e} = K_N {}^* K_N \qquad (12)$$

The nuclear kernels can generally be referred to as quarks.

From Eq. (7) and Eq. (12) both atomic states and nuclear states have action that is an integral fraction of the free electron action S_e.

The energy of the nuclear states in Eq. (12) is a deficit of energy and discussed in Appendix IV.

Multiple Kernel Nuclear Vertex Function

The atomic vertex kernels are the Rydberg states, and generally there are only two kernels. The nuclear kernels can bind inside the nucleus in any number to form complex particles, including atomic vertex kernels, as will be illustrated in the paper on vertex functions. When bound together, this becomes a Vertex Function of the Feynman path integrals.

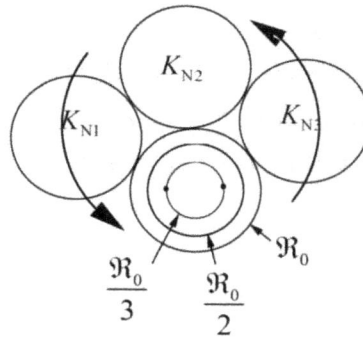

Multiple bound kernels of a nuclear vertex function

Fig. 5

There are generally a number of vertex kernels in the vertex function. They are each polarized and bind in pairs of opposite particles. Once bound, they have no polarization and are not aligned with other pairs or with external B fields.

Any group of vertex neutral pairs can be factored into two products, one with a positive kernel and one with a negative kernel, thus implying the effect of a positive and negative charge, as shown in Eq. (14).

$$K_P^* K_P = \left(K_1^* K_1\right)\left(K_2^* K_2\right)\left(K_3^* K_3\right) \cdot\cdot \quad \text{a.}$$

$$K_P^* = K_1^* \sqrt{\left(K_2^* K_2\right)\left(K_3^* K_3\right)\cdots} \quad \text{b.} \qquad (13)$$

$$K_P = K_1 \sqrt{\left(K_2^* K_2\right)\left(K_3^* K_3\right)\cdots} \quad \text{c.}$$

Eq. (14) a. shows a neutral vertex function as a group of un-aligned kernel pairs, and Eq. (14) b and c. shows positive and negative particles having a single polarized kernel.

Specific Particle Illustrations

Positronium Atomic States

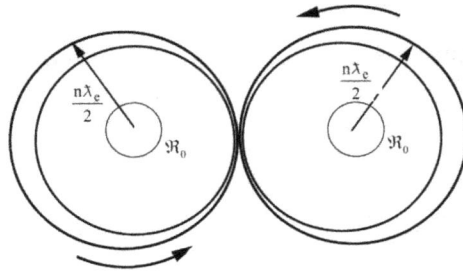

Fig. 5A

Noting the center of mass orbiting radius of the $\pm$ electrons in positronium is half the Compton radius of the electron, the action of Eq. (10) is reduced by half, but the integer is not changed. This is the proper interaction energy of a positive and negative electron establishing the Rydberg energy levels without the concept of charge, and is the exact energy levels of positronium: 6.803ev.

$$\left(K_A * K_A\right) = \left(\frac{S_e/2}{n_R \hbar g_A^2}\right)^* \left(\frac{S_e/2}{n_R \hbar g_A^2}\right) = \frac{\Delta\varepsilon}{m_e} \rightarrow \left(K_A * K_A\right) = 6.803\text{ev}/n_R$$

$$(14)$$

Muon Nuclear States

$$\begin{pmatrix} \dfrac{m_\mu}{m_e}^* & \dfrac{m_\mu}{m_e} \end{pmatrix} = \left[\left(\dfrac{\hbar}{S}\dfrac{\eta}{2^4} \right)^* \left(\dfrac{\hbar}{S}\dfrac{\eta}{2^4} \right) \right] \left[\left(\dfrac{\hbar}{S}\dfrac{\eta}{2^3} \right)^* \left(\dfrac{\hbar}{S}\dfrac{\eta}{2^4} \right) \right] \quad (15)$$

The muon is the lowest nuclear mass level, and Eq. (17) is the binding energy of the positive and negative conjugates of its vertex function. Factoring gives the vertex of the positive * and negative muons, and the positive and negative mass factors of the muon vertex are:

$$+\dfrac{m_\mu}{m_e} = \left(\dfrac{\hbar}{S_e}\dfrac{\eta}{2^4} \right) \left(\sqrt{\left(\dfrac{\hbar}{S_e}\dfrac{\eta}{2^3} \right)^* \left(\dfrac{\hbar}{S_e}\dfrac{\eta}{2^4} \right)} \right) = 206.7674344$$

$$-\dfrac{m_\mu}{m_e} = \left(\dfrac{\hbar}{S_e}\dfrac{\eta}{2^4} \right)^* \left(\sqrt{\left(\dfrac{\hbar}{S_e}\dfrac{\eta}{2^3} \right)^* \left(\dfrac{\hbar}{S_e}\dfrac{\eta}{2^4} \right)} \right) = 206.7674344$$

$$(16)$$

Particle Mass in Electrons

	Calculated from Theory		Experimental, Codata, & PDG
	Mass	GeV	Mass
Proton	1836.15267344		1836.15267344
Neutron	1838.69296925		1838.68366173
Tauon	3477.18825		3477.135 +/- .136
Muon	206.76743435		206.76828298
Pion+/-	273.131906		273.1321133
Z Boson	177190.369	90.544	90.04 - 90.76 GeV
W Boson	157359.58	80.410	80.433 +/-.0009 GeV
Top Quark	336153.82	171.7225	171.77 +/- 0.38 GeV CER CMS

Vertex Functions and Calculated Mass Values

The vertex functions of several primary particles have been discovered and are presented below with the calculated values. The fundamental constants as well as the mass values used in the calculations are included in Appendix II, from Codata, and the Particle Data Groups.

Particles Calculations For:

Proton, Neutron Tauon, Muon, Charged Pion, Z Boson, W Boson, Top Quark

Summary

The Discovered Particle Vertex Functions and the Calculated Rest Mass

Proton
$$\frac{\text{Proton}}{1836.15267344} \qquad \pm\frac{m_P}{m_e} = \sqrt{\left(\frac{\hbar}{S_e}\frac{2^7}{\eta^3}\right)\left(\frac{\hbar}{S_e}\frac{2}{\eta^3}\right)^{1/2}\left(\frac{\hbar}{S_e}\frac{1}{\eta^3}\right)^{1/2}} \qquad (17)$$

Neutron
$$\frac{\text{Neutron}}{1838.692969} \qquad \left(\frac{m_N}{m_e}-1\right) = \sqrt{\left(\frac{S_e}{\hbar}\frac{1}{\eta^3}\right)\left(\frac{\hbar}{S_e}\frac{g_A^3}{1}\right)}\sqrt{\left(\frac{\hbar}{S_e}\frac{2^7}{\eta^3}\right)\left(\frac{\hbar}{S_e}\frac{2}{\eta^3}\right)^{1/2}\left(\frac{\hbar}{S_e}\frac{1}{\eta^3}\right)^{1/2}} \qquad (18)$$

Muon
$$\frac{\text{Muon}}{206.76743435} \qquad \pm\frac{m_\mu}{m_e} = \left(\frac{\hbar}{S_e}\frac{\eta}{2^7}\right)^*\left(\sqrt{\left(\frac{\hbar}{S_e}\frac{\eta}{2}\right)^*\left(\frac{\hbar}{S_e}\frac{\eta}{1}\right)}\right) \qquad (19)$$

Tauon
$$\frac{}{3477.18825} \quad \left(+\frac{m_\tau}{m_e}\right) = \left(\frac{\hbar}{S/2}\frac{1}{\eta^3}\right)^* \sqrt{\left(\frac{\hbar}{S/2}\frac{1}{\eta^3}\right)^{1/4} \left(\frac{\hbar}{S}\frac{1}{\eta^3}\right)^{3/4}} = \sqrt{\left(\frac{1}{2^3}\frac{m_P}{m_e}\right)^3} \qquad (20)$$

Charged Pion
$$\frac{}{273.131906} \quad \pm\frac{m_{\pi+}}{m_e} = \left(\frac{S_e}{\hbar}\frac{2}{g_A^2}\right)^* \left[\left(\frac{\hbar\eta}{S_e}\right)^* \left(\frac{\hbar\eta}{S_e}\right)\right] = 2\alpha\left(\frac{\hbar\eta}{S_e}\right)^2 \qquad (21)$$

+ W Boson
$$\frac{}{177190.369} \quad \pm\frac{m_W}{m_W} = \sqrt{\left(\frac{\hbar}{S_e}\frac{2^5}{\eta^2}\right)\left(\frac{\hbar}{S_e}\frac{2^6}{\eta^2}\right)^{1/2}\left(\frac{\hbar}{S_e}\frac{1}{\eta^2}\right)^{1/4}\left(\frac{\hbar}{S_e}\frac{2}{\eta^2}\right)^{1/4}} \qquad (22)$$
$$\frac{}{80.410\ \text{GeV}}$$

Z Boson
$$\frac{}{177190.369} \quad \frac{m_Z}{m_e} == \sqrt{\left(\frac{\hbar}{\eta^3 S}\right)^2\left(\frac{\hbar}{S_e}\frac{2^6}{\eta^3}\right) \times \left(\frac{\hbar}{S_e}\frac{2}{\eta^3}\right)^{1/2}\left(\frac{\hbar}{S_e}\frac{1}{\eta^3}\right)^{1/2}} \qquad (23)$$
$$\frac{}{90.544\ \text{GeV}}$$

Top Quark
$$\frac{}{336153.82} \quad \frac{m_{TOP}}{m_{TOP}} = \left(\frac{m_{TOP}}{m_e}\right) = \left(\frac{\hbar}{S_e/3}\frac{\eta^3}{}\right)\left(\frac{\hbar}{S_e/6}\frac{\eta^3}{}\right) \qquad (24)$$
$$\frac{}{171.7225\,\text{GeV}}$$

All the vertex kernels that compose the particle mass are integral values of the action, and all but two have the cube of the Anomalous Nuclear ratio, η^3; the two are the W boson, η^2, and the charged pion, η. The exponents are related to the nuclear contents and configuration, but at this time that is not clear.

Some Connections and Discussion

Some of the relations between internal parts of nuclear particles seem obvious, some are obscure.

The product of the proton times its conjugate is two pair of conjugate quarks:

Proton
$$\left(\frac{m_P}{m_e}^* \frac{m_P}{m_e} \right) = \sqrt{\left(\frac{\hbar}{S_e} \frac{2^7}{\eta^3} \right)^* \left(\frac{\hbar}{S_e} \frac{2^7}{\eta^3} \right)} \sqrt{\left(\frac{\hbar}{S_e} \frac{2}{\eta^3} \right)^* \left(\frac{\hbar}{S_e} \frac{1}{\eta^3} \right)} \quad (25)$$

$$= \left(\frac{\hbar}{S_e} \frac{2^7}{\eta^3} \right) \left(\frac{\hbar}{S_e} \frac{2}{\eta^3} \right)^{1/2*} \left(\frac{\hbar}{S_e} \frac{1}{\eta^3} \right)^{1/2}$$

Factoring the conjugate positive and negative proton gives the individual positive and negative particles and can be factored into two, three quark particles:

$$+\left(\frac{m_P}{m_e} \right) = \sqrt{\left(\frac{\hbar}{S_e} \frac{2^7}{\eta^3} \right)^*} \left[\sqrt{\left(\frac{\hbar}{S_e} \frac{2}{\eta^3} \right)^{*}}^{1/2} \sqrt{\left(\frac{\hbar}{S_e} \frac{1}{\eta^3} \right)}^{1/2} \right] = 1836.15267344$$

$$(26)$$

$$-\left(\frac{m_P}{m_e} \right) = \sqrt{\left(\frac{\hbar}{S_e} \frac{2^7}{\eta^3} \right)} \left[\sqrt{\left(\frac{\hbar}{S_e} \frac{2}{\eta^3} \right)^{*}}^{1/2} \sqrt{\left(\frac{\hbar}{S_e} \frac{1}{\eta^3} \right)}^{1/2} \right] = 1836.15267344$$

$$\frac{m_P}{m_e} = \frac{m_1}{m_e} \frac{m_2}{m_e} \frac{m_3}{m_e} \quad (27)$$

The individual quark masses in electrons are:
$$132.1685537$$
$$3.4179175$$
$$4.0646118$$
And the product of these is: 1836.1526734 Electrons

These are not the same quark mass as measured by the Particle Data Group for the quarks which are 4.50 & 9.39 electrons.

Muon

$$\pm \frac{m_\mu}{m_e} = \left(\frac{\hbar}{S_e} \frac{\eta}{2^7} \right)^* \left(\frac{\hbar}{S_e} \frac{\eta}{2} \right)^{1/2*} \left(\frac{\hbar}{S_e} \frac{\eta}{1} \right)^{1/2}$$

The Muon has identical but reciprocal Rydberg integers of the proton-proton conjugate pair, Eq. (25).

Neutron

The neutron without the mass of the free electron is the product of the proton and a ratio of nuclear and atomic Kernels. This is presumably the electron neutrino and calculates out to be an energy of 1.000833800924 electrons. [21]

$$\left(\frac{m_N}{m_e}-1\right)=\sqrt{\left(\frac{S_e}{\hbar}\frac{1}{\eta^3}\right)\left(\frac{\hbar}{S_e}\frac{g_A^3}{1}\right)}\sqrt{\left(\frac{\hbar}{S_e}\frac{2^7}{\eta^3}\right)\left(\frac{\hbar}{S_e}\frac{2}{\eta^3}\right)^{1/2}\left(\frac{\hbar}{S_e}\frac{1}{\eta^3}\right)^{1/2}}$$

$$\frac{m_N}{m_e}=\sqrt{\left(\frac{S_e}{\hbar}\frac{1}{\eta^3}\right)\left(\frac{\hbar}{S_e}\frac{g_A^3}{1}\right)}\times\frac{m_P}{m_e}-\frac{m_e}{m_e}$$

The value $\sqrt{\left(\frac{S_e}{\hbar}\frac{1}{\eta^3}\right)\left(\frac{\hbar}{S_e}\frac{g_A^3}{1}\right)}$ = should be the binding energy of the electron with the proton and is the energy of the escaping neutrino.

Calculating this out gives a small undefined error.

Subtracting the two energy values in electrons gives the difference:

1.00083665196259 - 1.000833800924 = 00000285104 = 1.456 eV

This discrepancy is about twice the 0.8eV upper limit of the current KATRIN Electron Neutrino Experiment for the rest mass of the electron neutrino. The discrepancy is not necessarily connected to the neutrino rest mass, however.

Charged pion

$$\left(\frac{m_{\pi+}}{m_e}\right)=\left(\frac{S_e}{\hbar}\frac{1}{g_A^2}\right)^*\left[\left(\frac{\hbar\eta}{S_e/2}\right)^*\left(\frac{\hbar\eta}{S_e}\right)\right]=\alpha\left(\frac{\hbar\eta}{S_e/2}\right)^*\left(\frac{\hbar\eta}{S_e}\right)$$

The charged pion is an electron Kernel in the Compton orbit bound with two conjugate quarks.

Top quark

The top quarks appears to be the binding of two quarks,

$$\left(\frac{m_{TOP}}{m_e}\right)=\left(\frac{\hbar}{S_e/3}\frac{\eta^3}{}\right)\left(\frac{\hbar}{S_e/6}\frac{\eta^3}{}\right)$$

Conclusion

The theory presented here provides both the origin of charge and the mass of elementary particles. There is no numerology, approximates, or coupling constants. The theory defines the mass of particles in a consistent, logical, and well-known, but extended Feynman theory of Quantum Electrodynamics.

It was surmised by this author that nuclear particles would be defined by $SU(2)SO(3)$ group math structure. This turned out not to be the case, and the Dirac matrix of Geometrical Algebra combined with the action flow constructs of Feynman is sufficient to understand the mechanics.

The proof of concept of this theory is in the mass calculations shown in this paper.

References

DT Froedge, *The Physics of Delta-c Mechanics*, ISBN-13: 979-8218347178 (Feb. 14, 2024), (Papers can be found as chapters in this publication), https://www.amazon.com/Physics-Delta-C-Mechanics-Approach-Particle/dp/B0CVZ8CNYQ.

1. DT Froedge, "The Concepts and Principles of Delta-c Mechanics," August 2024, DOI: 10.13140/RG.2.2.31630.37445, https://www.researchgate.net/publication/382850589.

2. DT Froedge, "The Connection between Electric Charge, Gravitation, and the Feynman Sum over All Histories View of Quantum Electrodynamics," April 2020 Conference: APS, April 18–21, 2020 Washington, DC, https://absuploads.aps.org/presentation.cfm?pid=18355, https://www.researchgate.net/publication/341310206.

3. DT Froedge, "The Electron as a Composition of Two Vacuum Polarization Confined Photons," April 2021, DOI: 10.13140/RG.2.2.18971.18722, https://www.researchgate.net/publication/350740864.

4. DT Froedge, "The Fine Structure Constant from the Feynman Path Integrals," March 2021, DOI:10.13140/RG.2.2.12979.55846, https://www.researchgate.net/publication/350188862.

5. DT Froedge, "Vacuum Polarization, Gravitation, Charge, and the Speed of Light," Sept. 2021, DOI:10.13140/RG.2.2.15619.22569, https://www.researchgate.net/publication/354474157 (Equations 20–33).

6. DT Froedge, "The Calculated value of the Fine Structure Constant from Fundamental Constants," September 2021, DOI:10.13140/RG.2.2.34349.41440.

7. "CODATA values of the fundamental physical constants," https://www.nist.gov/programs-projects/codata-values-fundamental-physical-constants.

8. DT Froedge, "Structure of Elementary Nuclear Particles in Delta-c Mechanics," November 2023, DOI: 10.13140/RG.2.2.27181.70884, https://www.researchgate.net/publication/385782943.

9. CMS collaboration, "A profile likelihood approach to measure the top quark mass in the lepton + jets channel," April, 2022, https://home.cern/news/news/physics/cms-measures-mass-top-quark-unparalleled-accuracy.

10. M. Khodaverdian, "Accuracy and Precision of the Z Boson Mass Measurement with the ATLAS Detector," May 27, 2019, https://indico.cern.ch/event/813935/contributions/3557802/attachments/1919010/3174010/Gymnasieprojekt_Mariam_Khodaverdian_2019.pdf.

11. CDF Collaboration, "High-precision measurement of the W boson mass with the CDF II detector," April 2022. DOI: 10.1126/science.abk1781, https://www.science.org/doi/10.1126/science.abk1781.

12. DT Froedge, "The Dirac Equation and the two Photon Model of the Electron revised," April 2021, DOI: 10.13140/RG.2.2.19095.70564, www.researchgate.net/publication/350922403.

13. Arthur Eddington, *Relativity Theory of Protons and Electrons*, New York: The MacMillan Company Cambridge England, The University Press, 1936.

14. He et al., "Optomechanical measurement of photon spin angular momentum and optical torque in integrated photonic devices," Sept. 9, 2016, *Science Advances* Vol. 2, no. 9, DOI: 10.1126/sciadv.1600485.

15. DT Froedge, Feynman Vertex Functions, Researchgate, June 2025,DOI: 10.13140/RG.2.2.35247.24487, https://www.researchgate .net/publication/392331016

Recent Papers

The following recent papers by the author focus on the application, mechanism, and relation to current theory, as well as physical applications:

1a. DT Froedge, "Calculated Atomic Masses in Delta-c Mechanics," January 2025, DOI: 10.13140/RG.2.2.30292.51842, https://www .researchgate.net/publication/388219158.

2a. DT Froedge, "Structure of Elementary Nuclear Particles in Delta-c Mechanics," November 2024, DOI: 10.13140/RG.2.2.27181.70884, https://www.researchgate.net/publication/385782943.

3a. DT Froedge, "The Concepts and Principles of Delta-c Mechanics," August 2024, DOI: 10.13140/RG.2.2.31630.37445, www.researchgate .net/publication/382850589.

4a. DT Froedge, "Acceleration Gravitation and Origin of Centrifugal Force in Delta-c Mechanics," July 2024, DOI: 10.13140/ RG.2.2.28944.42247, https://www.researchgate.net/publication /381915343.

5a. DT Froedge, "Relativistic Time Dilation Illustrated in Delta-c Mechanics, The Twin Paradox is an Illusion," December 2023, DOI: 10.13140/RG.2.2.15584.46085, www.researchgate.net /publication/376831390.

5a. DT Froedge, "Feynman Flow Density Alternative to Wavefunctions," June 25, DOI:10.13140/RG.2.2.11759.14242, https:/ /www.researchgate.net/publication/392330748.

Appendix I

The Physical Constants and Particle Mass Values from Codata and the Particle Data Group Used in This Paper

Particle	mass (gms)	Mass (electrons)	Compton Radius (cm)
Electron	9.10938370150000E-28	1.0000000000	3.86159267943989E-11
Muon	1.88353162700000E-25	206.7682829838	1.86759430591295E-13
Proton	1.67262192369000E-24	1836.1526734400	2.10308910326354E-14
Tauon	3.16754000000000E-24	3477.1350000000	1.11055036503721E-14
Neutal Pion	2.40618001661378E-25	264.1430085130	1.46193257250244E-13
Chgd Pion	2.48806819637887E-25	273.1321133000	1.41381693102776E-13
Neutron	1.67492749804000E-24	1838.6836617325	2.10019415509539E-14
Wboson	1.43344876621262E-22	157359.5770234800	2.45399279311973E-16
Zboson	1.61700283014969E-22	177190.3690000000	2.17542781991699E-16

Physical Constants

$\bar{\lambda}_{PL}$	Planck Radius	1.61640095996445E-33	(cm)
α	Fine Structure Constant	7.29735253594845E-03	
$\hbar$	Planck Constant	1.05457181760000E-27	(cgs)
c_0	Velocity of light	2.99792458000000E+10	(cm/sec)
ν_e	Electron Freq.	1.23558996386000E+20	(hz)
m_e	Electron Mass eV	5.10998902000000E+05	(ev)
G	Gravitational Constant	6.67550533180000E-08	$cm^3/gm\,sec^2$

Composit Constnts

$\bar{\nu}_e$	Electron cycle number	1.23558996386000E+20	Eq 4
E_0	Nuc Atom Ground State	2.67493983646500E-05	Eq 5
$\mathfrak{R}_0$	ECR Radius	2.82447977709503E-13	(cm)Eq 3
g_A	Anom Gyromagnetic Ratio	1.00115965218073	eq 11
η	Nuclear Anomalous Spin	1.00060014721177	Eq 12

Appendix II

Anomalous QED Ratios

There are two anomalous ratios that play a role in nuclear-atomic systems, shown in Eq. (27) and Eq. (28), and these are the factors g_A and η. They are the QED-determined loop integral delay for particles moving in bound orbits. The ratio has no effect on the free electron, but when an electron is bound in rotation in the Feynman action, integral loops cause a delay that alters the particle rotation.

The first ratios that apply to atomic or Rydberg levels is the Anomalous Gyromagnetic ratio g_A. When an electron is moving in free space, there are no loop delays; however, when it is in rotation in the Compton orbit, the QED loop delays increase the radius, due to the QED time delay, and lower the binding energy. Its value has been theoretically predicted and measured to about twelve significant digits. The value is:

$$g_A = 1.0011596521807 \tag{28}$$

This factor is an offset of the Rydberg atomic levels.

The second is designated as Eta.η. It represents the QED effect on the rotating photons inside a nuclear vertex function.

$$\eta = 1.00060014721177 \qquad (29)$$

The value of η has not yet been determined by QED methods, but it should not be dismissed as arbitrary, as it is small but crucial to the precise mass of several unconnected known particles. [9] If there is a fudge factor in this development, this is it. This ratio, like g_A, is not calculated from physical constants.

Appendix III

The Rydberg Integer $n_R \, \hbar$ in

State Functions Rationalization

Solutions to bound kernels all require integral values $\hbar$ results from the rotating frequency of the kernels being proportional to $\hbar$. If the numbers are integers, then after a period of time the cycles repeat, whereas if they were irrational that could not be true.

Particles bind at the rotating most probable action radius, and the relative rotation of each bound particle must be an integral number of $n\hbar$, otherwise the binding cannot be stable (see Fig. 5).

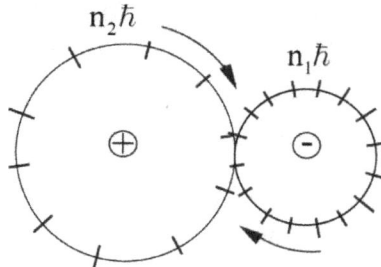

Fig. 6

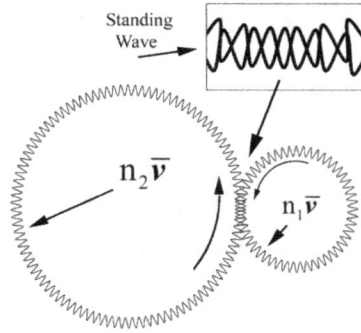

Fig. 5

The interaction of a kernel having action as $n_1\hbar$ with a kernel having action $n_2\hbar$, results in a standing wave of probability flow at the contact point that repeats as the particles counter-rotate. [18] A standing wave is a reciprocal flow probability between kernels that satisfies a unity continuity flow condition.

Note that a vertex function, such as Eq. (30), is not a legitimate kernel:

$$K \neq \left(\frac{\hbar}{S} \frac{\eta}{\sqrt{2}} \right)$$

(30)

The irrational number $\sqrt{2}$ cannot be in synchronization with integral multiples of $\hbar$. The mass value of this would be the same however if it were composed of two vertex kernels as shown Eq. (31):

$$K = \sqrt{\left(\frac{\hbar}{S} \frac{\eta}{2} \right)^*} \sqrt{\left(\frac{\hbar}{S} \frac{\eta}{1} \right)}$$

(31)

The mass of a vertex of a Feynman vertex Kernel is the product of an infinite number of products of action paths and, thus, the factors of Eq. (31) can be vertex kernels. Each of these vertices can bind to produce a neutral mass Kernel.

Appendix IV

Energy Deficit

The conjugate bound pairs in the nucleus are opposite and, if in free space, would annihilate. In binding, however, the electrons, like their atomic counterparts, have given up energy.

And, like the atomic counterparts, also have insufficient energy to escape. For a hydrogen atom, the state energy has to be kinetically injected to remove the electron. For a nuclear mass, the same principle applies, the energy of the entire mass has to be injected to remove the particle from the nucleus. This implies that nuclear mass exists as a deficit of energy, and particles in the nucleus cannot annihilate unless energetically removed.

This gives another perspective to the Einstein energy relation, in that nuclear mass exists as deficit of energy.

$$mc^2 = \varepsilon \quad \rightarrow \quad mc^2 = -\varepsilon \quad (32)$$

Calculated Atomic Masses in Delta-c Mechanics

D.T. Froedge
V020324

Abstract

The purpose of this paper is to illustrate the capability of Delta-c Mechanics and demonstrate results in nuclear structure unattainable by standard Newtonian physics, including the methodology of the Standard Model.

Delta-c Mechanics is a new formulation of physics with a broad range of applicability. While the author's recent papers [1, 2, 23] have primarily focused on applying the theory to problems already addressed by standard physics, yielding results that, though derived through the novel methods, are not new. This work seeks to highlight its unique potential. Specifically, to show how Delta-c Mechanics provides a clear and definitive framework for nuclear particle mass, with capability that has no equivalent in Newtonian or QM physics.

In Arthur Eddington's book from 1936 on photons and electrons, he observed that the wavefunctions of particles are multiplicative rather than additive. Based on this insight, he suggested that mass, like wavefunctions, might also arise as a product of its components rather than as their sum.

Delta-c Mechanics offers a framework where the relative masses of nuclear particles are entirely determined by the product of their constituent masses. This approach aligns with Eddington's view and provides a rigorous, mathematically grounded method for understanding mass relationships at a fundamental level.

By focusing on multiplicative principles, Delta-c Mechanics reveals new insights into the nature of mass and its role in particle physics, setting it apart from conventional theory.

Feynman's formulation of QED is based on the principle that particles take all possible action paths from vertex to vertex by summing amplitudes over all possible paths and obtaining the probability amplitude of the path.

Particles are the vertex of bound states and are poles in QED amplitudes. At the poles, the path integrals fail and are replaced with boundary conditions that match particle properties. The point at the vertex of the Feynman has conservation rules regarding the energy and particles in and out but does not define the interior mechanisms of the vertex.

Delta-c mechanics can define the interacting structure of the particle regarding the mass and probability density generated at the vertex and, thus, is able to define the relative mass of nuclear particles.

Topics

Vertex Functions

The structure of the vertex identified as a Vertex Function is composed of a product of a number of vertex kernels. A Vertex Kernel is the structure of the interacting probability density of electrons or neutrinos that can exist in various state conditions in units of Planck's constant.

The "Vertex Function" is the product of a collection Vertex Kernels that make up the mass of nuclear particles. The Vertex Kernels could be said to be sub-particles of a composite particle that define the probability density of the location of a sub-particle as a function of the distance from that particle. Originally this was developed in [8] for two electrons, it is clear now that it has general application for a multiplicity of nuclear particles.

With particles defined in Kernels, and the Vertex Function defined for a number of particles, the decay components can be identified, including that of the neutrino.

The proposed structure of the neutrino is developed in Part Two of this paper.

Structure of Elementary Nuclear Particles

In previous research, the mass values of particles within the atomic nucleus have been presented; however, this paper aims to delve deeper by exploring the internal structure of these particles and elucidating why elementary mass particles exhibit specific, quantifiable values. The foundational studies, *"Nuclear Particle Structure in Delta-C Mechanics"* and *"Neutrino Binding Between Nuclear Particles in Delta-C Mechanics"* [13, 27], laid the groundwork for this ongoing investigation, providing the initial direction for understanding nuclear particle structure. [2]

Preliminaries

Photons and Particles

Delta-c Mechanics postulates the photon is a rotating Planck particle having a radius of λ_{PL} (10E-33 cm), and a probability flow direction rotating at the Compton frequency, having energy of $\varepsilon = \hbar\omega$. If the photon is not in rotation, it has no energy, but still has an interaction with the probability flow of other photons by virtue of its cross section, changing the flow direction, thus having an effect on the index of refraction and the probability flow of other photons.

The primary rest mass particle is the electron, formed by the self-binding of two photons that have sufficient probability flow to alter the index of refraction of the other, into its orbit. (see [5] for details). The change in the velocity of light near a rotating particle due to the probability of its photon flow density is given by:

$$\frac{\Delta c}{c_0} = \left(\frac{\lambda_{PL} \bar{v}_e}{r} \right) \tag{1}$$

This is the elemental Vertex Kernel that provides both the probability density and the cross section of an electron. This is also the change in the velocity of light at a distance r from the rotation of a Planck particle at the repetition rate of the electron.

The product of two Electron Vertex function yields the product of the probability density of the two particles and, in turn, the mass generated by the interaction.

$$\frac{\Delta c}{c_0} = \frac{\Delta \varepsilon}{\varepsilon_0} = \frac{1}{2} \left(\frac{\sqrt{2}\lambda_{PL}\bar{v}_e}{r} \right)\left(\frac{\sqrt{2}\lambda_{PL}\bar{v}_e}{r} \right) = \left(\frac{\lambda_{PL}\bar{v}_e}{r} \right)\left(\frac{\lambda_{PL}\bar{v}_e}{r} \right) \tag{2}$$

This relation shows the change in the velocity of light at one particle at a distance r from the other, and giving the interaction energy $\Delta \varepsilon$ as a result of the particle biding $\Delta \varepsilon$ is the deficit in energy created by the binding, and is radiated away when the particles have an integral $\hbar$ solution, $n\hbar = mcr$. The numerators in Eq (2) are identified as the Electron Creation Radius. [5] $\Re_0$ is the radius that the two photons in the electron orbit each other.

$$\Re_0 = \sqrt{2}\lambda_{PL}\overline{\nu}_e \quad = \quad 2.82447977709503\text{E}-13 \text{ cm} \quad (3)$$

This, as well as other constants, are used in Delta-c Mechanics; they are not arbitrary, but a composition of fundamental constants, all of which are included in Appendix II.

The radius, $\Re_0$, is composed of the Planck particle radius $\lambda_{PL} = \sqrt{G\hbar/c^3}$ and $\overline{\nu}_e$, which is termed the Electron Rotation number. $\overline{\nu}_e$ is the ratio of the velocity light travels in free space to the relative velocity of the photons, $\overline{v}_e$, in the creation orbit of the electron. It is "unitless," but its magnitude is equivalent to the Compton Electron Frequency ν_e.

$$\nu_e = \frac{c_0}{\lambda_e} \qquad \overline{\nu}_e = \frac{c_0}{v_e} \qquad \rightarrow \qquad \left|\nu_e\right| = \overline{\nu}_e \qquad (4)$$

The velocity v_e is the velocity of light for the bound photons in the core of the electron slowed down by the binding probability density of the other.

The relations in Eq. (4) are developed in [5].

The Nuclear Atomic Ground State

Stable solutions to particle interactions exist when the radius of a particles rotation is such that the angular momentum is an integral number of the action quantum $\hbar n$, or $\hbar/n$. The ground state of atomic and nuclear particles exists at the Compton radius of the free electron.

$$E_0 \quad = \quad \frac{1}{2}\left(\frac{\Re_0}{\lambda_e}\right)^2 = \frac{1}{2}\left(\frac{\Re_0 m_e c_0}{\hbar}\right)\left(\frac{\Re_0 m_e c_0}{\hbar}\right) \qquad (5)$$

The ratio: $\mathfrak{R}_0 m_e c_0 / \hbar = 1/136.71872288679$, is the ratio of the angular momentum of the orbiting photons in the electron creation radius that form the electron, to the Planck action quantum, $\hbar$. This is the actual physical angular momentum of the free electron, not the relative value of $\hbar/2$ assigned to Fermions.

$$\mathfrak{R}_0 \left[p_1 + p_2 \right] / \hbar \qquad\qquad p_1 + p_2 = m_e c_0 \qquad (6)$$

This is a fundamental constant, and all numbers are known to at least ten significant places. All integers used in this paper are multiples or divides of $\hbar$ and do not change the calculated accuracies. This state marks the ground state of both atomic and nuclear particles, and there are no arbitrary adjustable factors.

The value of the binding of two electrons, Eq. (5), one positive and the other negative, forms the ground state of the atomic-nuclear particles. The value of E_0 is the unitless ratio of the state energy of two $\pm$ bound electrons, reference to the mass of an electron.

The Rydberg energy ground state of atomic particles is not exactly at the radius λ_e, but because of the QED loop induced delay in the completion of the electron Compton radius of two orbiting electrons, the radius is greater by the factor of the Anomalous Gyromagnetic Ratio g_A (see Eq. (11)). The base state radius is at the Compton radius of the electron, but the gyromagnetic ratio, which is a delay of orbital completion due to the probability loops, increases the Rydberg base radius of atomic particles to $r = \lambda_e g_A^2$, which establishes the ground state of atomic particles.

When this radius is set as the solution to Eq. (2), the energy levels are the ionization energy of two opposite $\pm$ electrons, or twice the ionization energy of positronium.

$$\frac{E_0}{g_A^4} = \frac{1}{2}\left(\frac{\mathfrak{R}_0}{\lambda_e g_A^2} \right)\left(\frac{\mathfrak{R}_0}{\lambda_e g_A^2} \right) = \frac{1}{2}\alpha^2 \rightarrow 13.606 \text{ eV} / m_e \qquad (7)$$

The difference between this and the hydrogen ionization is the inclusion of the reduced mass in the hydrogen atom.

The Rydberg integer n_R, which represents integral values of the quantum of action $\hbar$, can be included in this expression, defining the atomic levels:

$$\frac{\Delta\varepsilon}{\varepsilon_0} = \frac{\alpha^2}{2} \rightarrow \frac{m_X}{m_e} = \left(\frac{\alpha^2}{2n_R^2}\right) \tag{8}$$

The energy radiated away when the electron occupies the states is $m_e\left(\alpha^2/2n_R^2\right)$, which is the reduction in the mass of the electron occupying this position.

Note from this relation The Fine Structure Constant is a composite constant, and can be evaluated with Eq. (3) and Eq. (7); the product of fundamental constant is not arbitrary.

$$\alpha = \left(\frac{\lambdabar_{PL}\bar{v}_0}{\lambdabar_e g_A^2}\right) = 1/137.035999710 \tag{9}$$

Since α is the most accurately known number in this group, this relation is used to calculate the value of the Planck particle radius, and is:

$$\lambdabar_{PL} = 1.61640095996445E-33 \text{ cm} \tag{10}$$

This function, Eq. (7), shown with the value of the Rydberg integer $n_R = 1$, is central to both atomic and nuclear particles. While hydrogen atomic states exhibit only two particles in the binding, nuclear particles demonstrate multiple bindings of particles with the similar Kernels. The multiplicity defined by the Rydberg integer, n_K in Eq. (7), is present in the structure of atomic particles as well as nuclear particles.

Structure of Nuclear Particles

It is proposed that all particles are composed of the bindings of functions designated as kernels; one being the Electron Kernel, K_A, for atomic bound particles, and the other the Nuclear Kernel, K_N, for nuclear bound particles.

The base state of all particles as discussed is E_0 that is the binding energy of two free electrons (Equation(7)). States that have energy less than E_0 are atomic and states that have energy values greater than E_0 are nuclear states (particles). The Compton radius of the Rydberg atomic ground levels is slightly larger than the free electron λ_e due to the QED induced rotational delay of the Anomalous Gyromagnetic ratio g_A, such that $\lambda_R = \lambda_e g_A^2$.

Anomalous Gyromagnetic Ratio g_A

There are two anomalous ratios that play a role in nuclear-atomic systems; particularly, the mass of the neutrinos. They are the QED-determined loop integral delay for particles moving in a circle. The ratio has no effect on the free electron, but when it is bound in rotation, the Feynman action integral loops cause a delay that alters the electrons Compton radius. For atomic or Rydberg levels with radii above the electron, at $r = \lambda_e g_A^2$, the Anomalous Gyromagnetic ratio g_A^2 increases the electron Compton radius. Its value is:

$$g_A = 1.0011596521807 \tag{11}$$

Anomalous Nuclear Ratio η

The second is a similar factor that applies to particles with Compton radii less than $\mathfrak{R}_0$, it is here designated as Eta.η. The value can be determined with precision by relations between the mass of several elementary particles, and it is:

$$\eta = 1.00060014721177 \tag{12}$$

For nuclear particles, with Compton radii less than $\mathfrak{R}_0$, it is found that the QED effect on the Compton radius η is about ½ that of the Anomalous Gyromagnetic ratio (i.e., $\eta^2 \approx g_A$).

The Anomalous Nuclear ratio Eta η represents the change in the Compton radius for a mass particle, $\lambda'_x = \lambda_x \eta$, bound within the nucleus. This factor is consistently present in the mass of all nuclear states. As the value of the Anomalous Gyromagnetic ratio is very precise, this constant is also very precise. Eta (η) is exact to at least twelve significant digits and uniform across all nuclear particles. The exponent of Eta $\eta^{\pm n}$ depends on the number of mass particles bound, and the sign depends on relative spin. Although the value of η has not yet been determined by QED methods, it should not be dismissed as arbitrary, as it is crucial to the precise mass of several interconnected known particles. This factor, like the Anomalous Gyromagnetic ratio, represents an offset to the nuclear ground state. It alters the probability density of the kernel. If it were not for these anomalous factors, the state values of atomic and nuclear particles would not exist. (The relation of η to the mass of the neutrino is shown in Part Two)

Part One

The Particle Kernels

The form of the functions in Eq. (8) are the Kernel of atomic and nuclear particles K, and are defined as:

$$K_A = \left(\frac{\lambda_{PL}}{\lambda_e g_A^2} \frac{\bar{v}_e}{n_R} \right) \qquad K_N = \left(\frac{\lambda_{PL}}{\lambda_e \eta} \frac{\bar{v}_e}{n_N} \right) \quad (13)$$

Electron Kernel   **Quark Kernel**

These are the core probability densities, or the two-photon electron, the product of which establishes the energy of the interaction or the value of the interaction particle mass. Note that the Rydberg integer n_R is a multiple of the action quantum, $\left(n_R \lambda = (n_R \hbar)/mc \right)$, and n_N plays the same role in a nuclear kernel n_N. The kernels are the same except for the QED effects. The value of η in the kernel can be $\eta^{\pm n}$, indicating an increase or decrease in the electron radius, presumably the result of the relative spin between different particles. The difference in the two kernels is the QED effects induced by their position and, thus, the core **of all particles is an electron.**

Binding Mass

The ratio of the binding energy of two free electrons to the binding of two x particles yields the x particles mass to the mass of the electron. The binding of two particles is always the binding of a particle and an antiparticle.

The reciprocal of the root of the binding energy gives the state energy or the mass of the particle in that state.

$$\frac{E_X}{E_0} = \frac{\frac{1}{2}\left(\frac{\Re_0}{\lambda_X}\right)^2}{\frac{1}{2}\left(\frac{\Re_0}{\lambda_e}\right)^2} \rightarrow \left(\frac{E_X}{E_e}\right) = \left(\frac{m_X}{m_e}\frac{m_X}{m_e}\right) \rightarrow \sqrt{\frac{E_X}{E_0}} = \frac{m_X}{m_e} \qquad (14)$$

E_0 is the ground state of the nuclear-atomic system. E_X is the state of atomic and nuclear particles referenced to the ground state. It is the square of the mass of the particle occupying the state.

The ratio in Equation (14) is the binding energy of the positive and negative x particle and base state relative to the ground state energy. The last term is the expression for the mass of the single x particle.

Structure of Vertex Particles

As atomic bindings create a state particle, namely photons, from the binding of two electrons, the binding of two or more sub-particles defined by the Kernels in Eq. (13) can create state particles at vertex locations. These are the particles that occupy the vertex positions in Feynman diagrams and will be referred to as Vertex Functions. The states of atomic particles are directly proportional to the product of the vertex kernels, and the nuclear states are inversely proportional.

$$\frac{E_X}{E_0} = \left[(K_{A1})(K_{A2})\right] \qquad\qquad \frac{E_X}{E_0} = \frac{1}{(K_{N1})(K_{N2})(K_{N3})}$$

$$\frac{m_X}{m_e} = \sqrt{\left[(K_{A1})(K_{A2})\right]} \qquad\qquad \frac{m_X}{m_e} = 1/\sqrt{(K_{N1})(K_{N2})(K_{N3})} \qquad (15)$$

Atomic Kernel   **Nuclear Kernel**

The magnitude of the difference in the atomic and nuclear kernels effect on the vertex function can be illustrated as:

$$\frac{m_X}{m_e} \approx \frac{\Re_0 m_e c_0}{\hbar\, n} \sim \frac{1}{136\,n} \qquad\qquad \frac{m_X}{m_e} \approx \frac{1}{\dfrac{\Re_0 m_e c_0}{\hbar n}} \sim \frac{136\,n}{1} \quad (16)$$

Atomic Kernel   **Nuclear Kernel**

The Atomic Kernel in the vertex function is the ratio of angular momentum of the electron at the Electron Creation radius and the action quantum $\hbar$. The Nuclear Kernel in the vertex function is the reciprocal of the atomic Kernel and is six orders of magnitude greater; thus, the nuclear masses are far larger than the reduced electrons of the atomic states.

For particles such as the pion or the free neutron, the vertex function are products of both atomic and nuclear kernels. Meaning that the atomic vertex represents particles in Compton radii (10e-11 cm), whereas the nuclear kernels are particles bound in radii on the order of the electron radius (10e-15 cm).

$$E_X = E_0 \frac{\left[(K_{A1})(K_{A2})\right]}{\left[(K_{N1})(K_{N2})(K_{N3})\right]} \qquad\qquad \frac{m_X}{m_e} = \sqrt{\frac{\left[(K_{A1})(K_{A2})(K_{A3})\right]}{\left[(K_{N1})(K_{N2})\right]}} \quad (17)$$

The difference in atomic and nuclear particles is that in the atomic case, the photons state value radiates away the energy, reducing the mass of the bound particle, but for a nuclear particle the energy radiated away exceeds the free particle rest mass and the bound particles are electrons trapped as deficits in energy. All sub-particles have the same kernel as electrons, with the difference being the QED factors g_A and η. The atomic binding energy of the K_A particles, Eq. (15), is directly proportional to the product of the kernels, whereas the binding energy of the K_N components is inversely proportional. Product of the two signifies combinations of nuclear and atomic bindings. (There is a discussion of deficit energy in Appendix III.)

The simplified result of the internal mass of the particles could be stated as:

$$\left(\frac{m_e}{m_X}\right)^2 = \left(\frac{m_e}{m_1}\right)^2 \left(\frac{m_e}{m_2}\right)^2 \left(\frac{m_e}{m_3}\right)^2 \qquad (18)$$

The mass of m_X is not the sum of the internal particle mass, but the product. In disintegration, these individual vertices can bind with others to form temporary particles.

Calculated Mass Values

The vertex functions of several primary particles have been discovered and are presented below; the accuracy of all, but the muon, is within experimental tolerance. The muon's calculated mass is outside the experimental tolerance by a yet unexplained one part in one-fourth of a million.

The fundamental constants, as well as the mass values used in the calculations, are included in Appendix II, from Codata and the Particle Data Groups.

The particles calculations illustrated are for:

Proton, Neutron, Tauon, Muon, Charged Pion, Z Boson,
W Boson, Top Quark

Summary

Particle Mass in Electrons

	Calculated from Theory		Experimental, Codata, & PDG	
	Mass	GeV	Mass	
Proton	1836.15267344		1836.15267344	
Neutron	1838.68366173		1838.6838346173	
Tauon	3477.18825		3477.135 +/- .136	
Muon	206.76743435		206.76828298	
Pion+/-	273.131906		273.1321133	
Z Boson	177190.369	90.544	90.04 - 90.76	GeV
W Boson	157359.58	80.410	80.433 +/-.0009	GeV
Top Quark	336153.82	171.7225	171.77 +/- 038	Gev

Particle Vertex Functions and Calculated Rest Mass

Proton
1836.15267344

$$\left(\frac{m_P}{m_e}\right) = 1 \Big/ \sqrt{\left(\frac{\lambda_{PL}\eta^3}{\lambda_e}\frac{\bar{v}}{2^6}\right)\left(\frac{\lambda_{PL}\eta^3}{\lambda_e}\frac{\bar{v}}{1}\right)^{1/2}\left(\frac{\lambda_{PL}\eta^3}{\lambda_e}\frac{\bar{v}}{2}\right)^{1/2}}$$

(19)

Neutron
1838.68366173

$$\left(\frac{m_N}{m_e}\right) = \sqrt{\frac{\left(\frac{\lambda_{PL}\bar{v}}{\lambda_e\eta^3}\right)}{\left(\frac{\lambda_{PL}\bar{v}}{\lambda_e g_A^3}\right)}} \frac{1}{\sqrt{\left(\frac{\lambda_{PL}\eta^3\bar{v}}{\lambda_e 2^6}\right)\left(\frac{\lambda_{PL}\eta^3\bar{v}}{\lambda_e}\right)^{1/2}\left(\frac{\lambda_{PL}\eta^3\bar{v}}{\lambda_e 2}\right)^{1/2}}}$$

(20)

Muon
206.76743435

$$\left(\frac{m_\mu}{m_e}\right) = 1 \Big/ \left[\frac{1}{\sqrt{2}}\left(\frac{\lambda_{PL}}{\lambda_e\eta}\frac{\bar{v}_e 2^4}{}\right)\left(\frac{\lambda_{PL}}{\lambda_e\eta}\frac{\bar{v}_e 2^4}{}\right)\right]$$

(21)

Tauon
3477.18825

$$\left(\frac{m_\tau}{m_e}\right) = 1 \Big/ \sqrt{\left(\frac{\lambda_{PL}}{\lambda_e}\frac{\bar{v}_e}{1}\frac{\eta^3}{}\right)^2\left(\frac{\lambda_{PL}}{\lambda_e}\frac{\bar{v}_e\eta^3}{1}\right)^{1/4}\left(\frac{\lambda_{PL}}{\lambda_e}\frac{\bar{v}_e}{2}\frac{\eta^3}{}\right)^{3/4}}$$

(22)

Charged Pion
273.131906

$$\left(\frac{m_{\pi+}}{m_e}\right) = \frac{\sqrt{2}\left(\frac{\lambda_{PL}}{\lambda_e g_A^2}\frac{\bar{v}_e}{1}\right)}{\left(\frac{\lambda_{PL}\bar{v}_e}{\lambda_e\eta^2}\right)^2} \rightarrow = \frac{\alpha}{\left(\frac{\lambda_{PL}\bar{v}}{\lambda_e\eta^2}\right)^2}$$

(23)

Z Boson
177190.369

$$\left(\frac{m_z}{m_e}\right) = \frac{1}{2}\times 1 \Big/ \sqrt{\left(\frac{\lambda_{PL}\eta^3}{\lambda_e}\frac{\bar{v}_e}{1}\right)^2\left[\left(\frac{\lambda_{PL}\eta^3}{\lambda_e}\frac{\bar{v}_e}{2^6}\right)\left(\frac{\lambda_{PL}\eta^3}{\lambda_e}\frac{\bar{v}_e}{}\right)^{1/2}\left(\frac{\lambda_{PL}\eta^3}{\lambda_e}\frac{\bar{v}_e}{2}\right)^{1/2}\right]}$$

90.544 GeV

$$\left(\frac{m_z}{m_e}\right) = \frac{1}{2}\left(\frac{m_P}{m_e}\right)\times 1 \Big/ \left(\frac{\lambda_{PL}\eta^3}{\lambda_e}\frac{\bar{v}_e}{1}\right)$$

(24)

W Boson

177190.369
80.410 GeV

$$\left(\frac{m_W}{m_e}\right) = \cfrac{1}{\sqrt{\left(\frac{\lambda_{PL}\bar{v}_e\eta^3}{\lambda_e 2^{11}}\right)\left(\frac{\lambda_{PL}\bar{v}_e\eta^3}{\lambda_e 2}\right)^{1/2}\left(\frac{\lambda_{PL}\bar{v}_e\eta^3}{\lambda_e 2}\right)^{1/4}\left(\frac{\lambda_{PL}\bar{v}_e\eta^3}{\lambda_e}\right)^{1/4}}} \qquad (25)$$

Top Quark

336153.82
171.7225 GeV

$$\left(\frac{m_{TOP}}{m_e}\right) = 1 \Big/ \left(\frac{\lambda_{PL}\eta}{\lambda_e}\frac{\bar{v}_e}{3}\right)^2 \qquad (26)$$

Summary—Part One

Without any injected coupling constants, the mass numbers are exceptionally accurate. Of the calculated mass values presented, only the muon mass is out of experimental tolerance by an unexplained one part in one-fourth of a million.

Earlier papers have focused on the physical aspects of Delta-c Mechanics. [1, 2] This paper's intention is to extend the foundational understanding of particle mass by focusing on the structure of the mass particles and the mechanical interactions.

Finding that nuclear particles inertial mass is created by a deficit of energy (Appendix III) produces an understanding of the balance of energy in the universe between electromagnetic and inertial mass: $-\varepsilon = mc_0^2$. A proton is actually the same as two positrons and an electron, trapped as an energy deficit of 1833.15 electrons.

Part Two

Neutrino Mechanical Structure

The mechanical structure of the neutrino described here is a bit speculative, but the vertex energy appears to be quite accurate.

The structure of the neutrino is similar to the electron, accept that in the electron the rotating bound photons move in opposite directions around a center of mass. The external Feynman photon probability density rotates in the same direction and thus provides particle interactions. For the neutrino, the photon density moves in the same direction with internal Compton rotational directions that are opposite but also rotate around the center of momentum. The exterior probability flow for the Compton rotation cancels the Compton radius, and particle-particle interaction.

In the paper on the Dirac equation [24], it was noted that for particle dynamics there are three parts of the energy: the inertial, the kinetic, and the total, that can be found as the magnitude of the four differentials. For the neutrino, the kinetic is the sum of the moving photons, and the inertial is the rotary energy of the photons about the center of momentum.

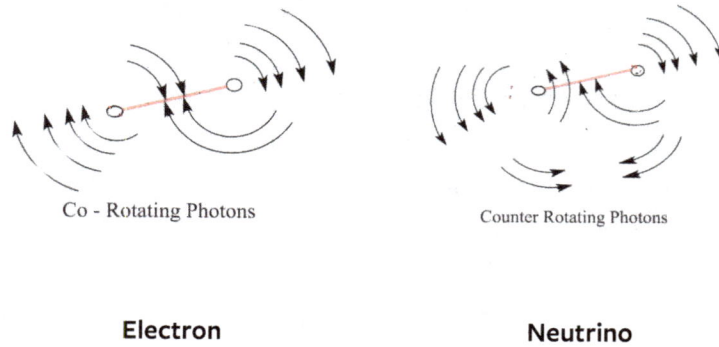

Co - Rotating Photons   Counter Rotating Photons

Electron   **Neutrino**

Fig. 1a.   **Fig. 1b.**

For the electron, photons rotate around a stationary center of mass (Fig. 1a.) bound by the reduction in the index of refraction induced by the mutual flow of their probability density. This interaction is planar, and the rotatory probability flow density is parallel to the velocity vector.

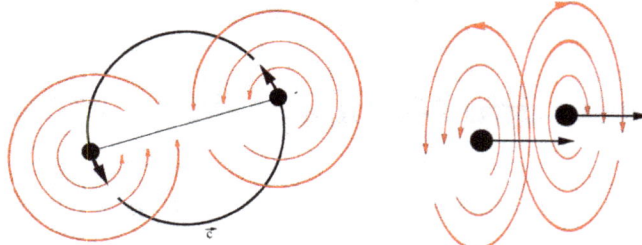

Fig 2a. Electron   **Fig 2b. Neutrino**

The neutrino is composed of two opposite rotating, photons, bound together with parallel helictical trajectories rotating around the center of momentum. The photons are rotating around each other, thus canceling the probability density flow exterior to the radius of rotation. The energy of the photons is equal to the two photons in the electron, thus the sum of the electromagnetic energy is equal to the mass of an electron.

The photons of the neutrino (Fig. 2b.) are moving linearly around a helical path at the velocity of light. The forward velocity is retarded by the path delay as the result of the rotational transverse component. The rotary motion component is an energy of the neutrino that is not altered by a velocity transform along the rotary axis, thus having the property of an inertial mass.

Neutrino Helictical Photon Paths

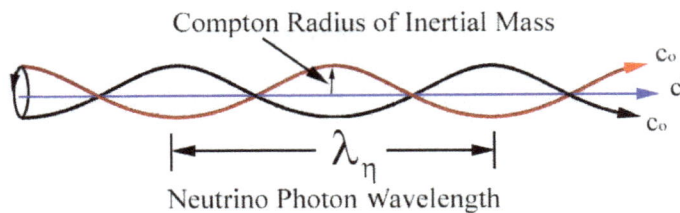

Compton Radius of Inertial Mass

λ_η

Neutrino Photon Wavelength

Fig. 3

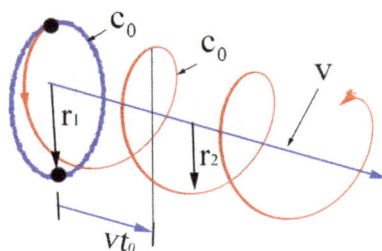

(showing only one photon)

Fig. 3A

69

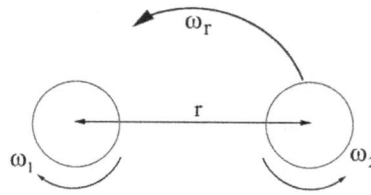

Fig. 4

Helictical Projections

The energy components an be graphically illustrated as the trajectory of the photons projected from the cylinder onto a plane, giving a pictoral view of the velocity mass relations Fig.5

Relativistic Moving Electron

$$m_e c_0^2 = m_x \left(c_0^2 + v^2 \right) \qquad m_x c_0^2 = \frac{m_e c_0^2}{\left(c_0^2 + v^2 \right)}$$

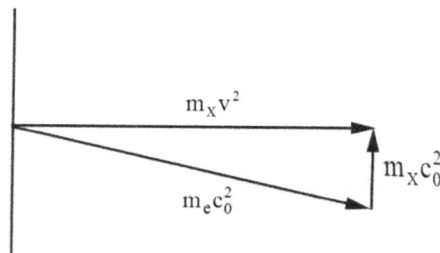

Fig. 5

70

The Neutrino Energy

For the rotating photons in the neutrino (Fig. 4), the revolving photons have a velocity of c_0, moving around a cylindrical trajectory. The velocity of the center of momentum defines its kinetic velocity, which is moving at a velocity less than c_0. The relationship between the components can be found by observing the frequency as it revolves.

The rotating photon, c_0, frequency is:

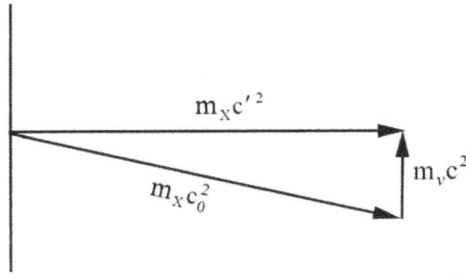

Fig. 6

$$\omega_x = \frac{c_0}{\lambdabar_n} = \frac{(m_x c_0) c_0}{\hbar} \tag{27}$$

The Kinetic frequency is:

$$\omega' = \frac{c'}{\lambdabar'_2} = \frac{(m_x c') c'}{\hbar} \tag{28}$$

The rotary frequency and energy is:

$$\omega_r = \frac{c}{\lambdabar_r} = \frac{(m_r c) c}{\hbar} = \frac{m_r c_0^2}{\hbar} \tag{29}$$

A Lorentz velocity transform along the center of momentum creates a Doppler change in the energy of the photons, but the rotating angular component does not change as a result of the motion. The rotary component is independent of the shift, and thus defines a constant energy component, or rest mass. It is presumed that mass is related to the rotating frequency $m_r = \hbar \omega_r$, and is the rest mass of the neutrino.

$$m_E \left(c_0^2 - c'^2 \right) = m_r c_0^2 \tag{30}$$

The angular frequency of the rotation of the photons around the axis matches the frequency of the linear projection on a plane, but the rotational radius and thus the rotational velocity is independent of the photon energy. The radius of the rotation is found by initial conditions, producing a rest mass of the neutrino. The total energy of the neutrino is developed in the section below on the vertex functions.

Vertex Energy of the Neutrinos

The Electron Neutrino from Neutron Decay

The neutron is an unstable particle composition of a proton, electron, and an anti-electron neutrino, and when it ejects these particles.

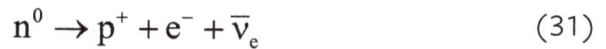

$$n^0 \rightarrow p^+ + e^- + \bar{v}_e \tag{31}$$

By observation of the electron decay when its kinetic energy is zero, the maximum energy of the emitted neutrino from the neutron can be determined, and thus the binding energy provided by the electron neutrino. [22]

From the excerpts of the paper on the details of nuclear particle structure [13], the vertex functions of both the neutron and the proton have been found. The proton vertex function Eq. (19) is composed of three quark kernels.

$$Proton\ Vertex\left(\frac{m_P}{m_e}\right) = \cfrac{1}{\sqrt{\left(\cfrac{\lambda_{PL}\eta^3}{\lambda_e}\cfrac{\overline{v}}{2^6}\right)\left(\cfrac{\lambda_{PL}\eta^3}{\lambda_e}\cfrac{\overline{v}}{1}\right)^{1/2}\left(\cfrac{\lambda_{PL}\eta^3}{\lambda_e}\cfrac{\overline{v}}{2}\right)^{1/2}}}\ *$$

(32)

This calculates to have a proton mass equal to:

$$m_P = 1836.15267344223\ electrons$$

*Note that $\overline{v}_e$ is the electron rotation number noted in Eq. (4), not the neutrino energy $\overline{V}_e$.

The vertex function for the Neutron from Eq. (20) is:

$$Neutron\ Vertex\ \left(\frac{m_N}{m_e}\right) = \underbrace{\cfrac{\sqrt{\left(\cfrac{\lambda_{PL}\overline{v}}{\lambda_e\eta^3}\right)}}{\sqrt{\left(\cfrac{\lambda_{PL}\overline{v}}{\lambda_{eA}g^3}\right)}}}_{\leftarrow\ Electron\ Neutrino\ Vertex\ Function}\cfrac{1}{\sqrt{\left(\cfrac{\lambda_{PL}\eta^3\overline{v}}{\lambda_e 2^6}\right)\left(\cfrac{\lambda_{PL}\eta^3\overline{v}}{\lambda_e}\right)^{1/2}\left(\cfrac{\lambda_{PL}\eta^3\overline{v}}{\lambda_e 2}\right)^{1/2}}}$$

(33)

Comparison shows the vertex function of the neutron contains the vertex function of the proton, and an additional two kernels that are the ratio of an electron and a quark Kernel. That term is thus the vertex function of the emitted neutrino without kinetic energy.

Calculated Neutron Mass

The additional two kernels in Eq. (33) constitute the ratio of an Electron Kernel and a Quark Kernel. Thus, the mass of the Neutron is the product of the mass of the proton and the included term. It is concluded that the Electron Neutrino Vertex Function is:

$$Neutrino\ Vertex\ Function\ \frac{\overline{v}_e}{m_e} = \sqrt{\cfrac{\left(\cfrac{\lambda_{PL}\overline{v}}{\lambda_e\eta^3}\right)}{\left(\cfrac{\lambda_{PL}\overline{v}}{\lambda_{eA}g^3}\right)}} = \frac{1.00083665196259\quad calc}{1.00083887131800\quad Exper}$$

(34)

The mass of the neutron is the binding energy and the mass of the electron:

$$\left(\frac{m_N}{m_e}\right) = \frac{m_P}{m_e} \times \frac{\overline{v}_e}{m_e} + m_e = \frac{m_P}{m_e} \times 1.000833802282540 + m_e \tag{35}$$

Adding the electron then gives **the mass of the neutron** within experimental error to be:

$$. \varepsilon_v - \varepsilon_C = 1.000836652 - 1.000838872 = 0.00000284967 \text{ electrons}$$

$$(36)$$

The difference between the experimental and calculated energy of the neutrino binding the electron to the proton in the neutron is:

$. \varepsilon_v - \varepsilon_C = 1.000836652 - 1.000838872 = 0.00000284967$ electrons   or $1.4561 eV$. The unexplained discrepancy is about 1 part in 350,000

Muon Neutrino

The muon neutrino can be found by its expulsion from the pion. The pion decays into a muon, and a muon neutrino.

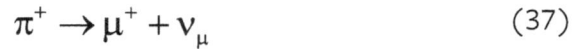

$$\pi^+ \rightarrow \mu^+ + v_\mu \tag{37}$$

The pion is thus a composition of the muon and a muon neutrino. It has a vertex function shown in Eq. (23), and the muon also has a vertex function as shown in Eq. (21).

Pion Vertex Function

The pion vertex structure at Eq. (23) is:

Pion Vertex
$$\left(\frac{m_{\pi+}}{m_e}\right) = \sqrt{2}\frac{\left(\dfrac{\lambda_{PL}\bar{v}_e}{\lambda_e g_A^2}\right)}{\left(\dfrac{\lambda_{PL}\bar{v}_e}{\lambda_e \eta}\right)^2} = \frac{\alpha}{\left(\dfrac{\lambda_{PL}\bar{v}_e}{\lambda_e \eta}\right)^2} \quad (38)$$

The value of the Fine Structure Constant in fundamental constants is:

$$\alpha = \left(\frac{\sqrt{2}\lambda_{PL}\bar{v}_e}{\lambda_e g_A^2}\right) \quad (39)$$

The vertex function of the Muon found in Eq. (21) is:

Muon Vertex
$$\left(\frac{m_\mu}{m_e}\right) = \frac{1}{\dfrac{1}{\sqrt{2}}\left(\dfrac{\lambda_{PL}\ \bar{v}_e 2^4}{\lambda_e \eta}\right)\left(\dfrac{\lambda_{PL}\ \bar{v}_e 2^4}{\lambda_e \eta}\right)} \quad (40)$$

Its decay products are:

$$\pi^+ \rightarrow \mu^+ + \nu_\mu \quad (41)$$

The vertex function of the muon Eq.(40) , can be factored be from the pion vertex function in Eq. (38), leaving:

$$\left(\frac{m_{\pi+}}{m_e}\right) = \left(\frac{m_\mu}{m_e}\right) \times \left(\frac{\lambda_{PL}\bar{v}_e 2^8}{\lambda_e g_A^2}\right) = \left(\frac{m_\mu}{m_e}\right)(1+.3209619104) \quad (42)$$

The remaining term is then the Muon neutrino. The mass contribution to the pion due to the neutrino is then:

$$\left(\frac{m_\mu}{m_e}\right) \times .3209619104 \; = 66.3647431366 \; \text{ electrons} \qquad (43)$$

When measured, this is reduced by: 8.062 electrons to 58.302 m_e due to the kinetic energy of the escaping muon. The value has been experimentally verified. [23]

The expelled neutrino vertex function, or the energy at rest in the pion, is:

$$\textit{Neutrino Expulsion Energy} \qquad \frac{v_\mu}{m_e} = \left(\frac{\lambda_{PL}\overline{v}_e}{\lambda_e g_A^2}\right) 2^8 = 675 \; \text{Kev} \qquad (44)$$

The energy of the muon neutrino expelled is 0.3209619104 electrons, or 628.12 Kev.

Since the factor of 2^8 originated in the pion and is not included in the muon vertex, it is presumed to be the kinetic energy ejected with the neutrino; thus, from the proposed model, the rest mass of the muon neutrino is the vertex in parenthesis in Eq. (44), and is :

$$v_{\mu 0} = 2.636 \; \text{KeV} \qquad (45)$$

Tauon Neutrino

The tauon particle can decay into a charged pion and a tau neutrino:

$$m_\tau \rightarrow m_{\pi+} + \overline{v}_\tau$$

76

There is a pion vertex function in Eq.(23) , and the pion in Eq. (38), thus the tauon neutrino vertex function can be arrived at by:

$$\left(\frac{m_\tau}{m_e}\right) = \left(\frac{m_{\pi+}}{m_e}\right)\bar{v}_\tau = \left(\frac{m_\mu}{m_e}\right) \times \bar{v}_\mu \bar{v}_\tau \qquad (46)$$

That gives the value of the tauon neutrino vertex function:

$$\left(\frac{m_\tau}{m_e}\right) = \frac{m_{\pi+}}{m_e}\bar{v}_\tau = \left(\frac{m_\mu}{m_e}\bar{v}_\mu\right) \times \bar{v}_\tau \quad \rightarrow \bar{v}_\tau = \left(\frac{m_\tau}{m_{\pi+}}\right) \qquad (47)$$

Tauon neutrino vertex energy is then:

$$\bar{v}_\tau = \left(\frac{m_\tau}{m_{\pi+}}\right) = 12.7308 \text{electrons.} \qquad 41.986 \text{ kev} \qquad (48)$$

After the Tau decays, only the pion remains; thus, the total ejection energy available is: $m_\tau - m_{\pi+} = 3204.$ electrons or 1.637 Gev

Sketches of the charged pion and the neutrino are shown in appendix IV Fig.2a and 3a.

Summary—Part Two

The neutrino vertex function represents energy of the neutrino's presence in a particle and increases the total mass of that particle. The total ejection energy of the neutrinos depends on the mass to which it is bound and the rest mass of the ejected particles. It should be, but it is not clear that this is the relativist mass of the neutrino when it is moving since it is not in agreement with ongoing experiments. [25]

The vertex energy/mass of the neutrino is the energy of two electron photons equal to the mass of the electron plus the rotational mass.

Presumably, the neutrino can be slowed down by particle interaction, but, at a minimum, its interaction with its antiparticle must produce a positron-electron. Therefore, its minimum energy is that of an electron.

Neutrino	Vertex Energy	Ejection Energy
Electron	426.023 eV	782.273 Kev
Muon	2.636 KeV	33.911 Mev
Tauon	41.986 kev	1.637 Gev

Conclusion

The particle mass numbers prove the concept. There are no arbitrary constants in the calculations, and if any values of the integers or the gyromagnetic ratios were off, changed by one part in a hundred thousand, the particle masses would be noticeably off in different directions. The agreement of the calculations with experimental values on so many particles proves the theory's merit.

Delta-c Mechanics is a new approach to physics within the framework of the Feynman path integral approach to QED. It has no fields, and no forces. Particles interact by the mutually induced a change in the index of refraction to the action path probability flow.

The methods of Delta-c Mechanics apply to all phenomena, including centrifugal force and gravitation, but standard theory produces equivalent results for most mechanical phenomena. The purpose of this paper has been to show accurate mass calculations for nuclear particles without the injection of arbitrary coupling constants, with results that are beyond question and cannot be produced by any existing theory.

The neutrino physical model is a bit speculative, but the vertex functions accurately produce the calculated mass of the neutron.

Author's Note

The material in Delta-c Mechanics, developed by the author over a number of years, is not for the casual observer. It is complicated and not likely to come into the physics community's notice for a long time. Eventually, however, it will find its way, for it is a new formulation and is the only theoretical approach that can produce the content of this paper.

References

DT Froedge, *The Physics of Delta-c Mechanics*, ISBN-13: 979-8218347178 (Feb. 14, 2024), (Papers 3–14 can be found as chapters in this publication), https://www.amazon.com/Physics-Delta-C-Mechanics-Approach-Particle/dp/B0CVZ8CNYQ.

1. DT Froedge, "The Concepts and Principles of Delta-c Mechanics," August 2024, DOI: 10.13140/RG.2.2.31630.37445, https://www.researchgate.net/publication/382850589.

2. DT Froedge, "Acceleration Gravitation and Origin of Centrifugal Force in Delta-c Mechanics," July 2024, DOI: 10.13140/RG.2.2.28944.42247, https://www.researchgate.net/publication/381915343.

3. DT Froedge, "The Connection between Electric Charge, Gravitation, and the Feynman Sum over All Histories View of Quantum Electrodynamics," April 2020 Conference: APS, April 18–21, 2020 Washington, DC, https://absuploads.aps.org/presentation.cfm?pid=18355, https://www.researchgate.net/publication/341310206.

4. DT Froedge, "A Quantum Theory Conjecture on the Origin of Gravitational and Electric Particle Interaction," December 2019, DOI: 10.13140/RG.2.2.29097.54884, https://www.researchgate.net/publication/337826826.

5. DT Froedge, "The Electron as a Composition of Two Vacuum Polarization Confined Photons," April 2021, DOI: 10.13140/RG.2.2.18971.18722, https://www.researchgate.net/publication/350740864.

6. DT Froedge, "The Gravitational Constant to Eleven Significant Digits," March 2020, DOI: 10.13140/RG.2.2.32159.38564, https://www.researchgate.net/publication/339943651.

7. DT Froedge, "The Fine Structure Constant from the Feynman Path Integrals," March 2021, DOI:10.13140/RG.2.2.12979.55846, https://www.researchgate.net/publication/350188862.

8. DT Froedge, "Vacuum Polarization, Gravitation, Charge, and the Speed of Light," Sept. 2021, DOI:10.13140/RG.2.2.15619.22569, https://www.researchgate.net/publication/354474157 (Equations 20–33).

9. DT Froedge, "The Calculated value of the Fine Structure Constant from Fundamental Constants," September 2021, DOI:10.13140/RG.2.2.34349.41440.

11. "CODATA values of the fundamental physical constants," https://www.nist.gov/programs-projects/codata-values-fundamental-physical-constants.

12. DT Froedge, "Electron Mass and State Energy Levels Resulting from Photon-Photon Interaction," Conference APS, April 2022, Conference: APS April 2022, New York, https://www.researchgate.net/publication/359912763.

13. DT Froedge, "Structure of Elementary Nuclear Particles in Delta-c Mechanics," November 2023, DOI: 10.13140/RG.2.2.27181.70884, https://www.researchgate.net/publication/385782943.

15. DT Froedge, "Neutrino Binding Between Nuclear Particles in Delta-c Mechanics," May 2024, DOI:10.13140/RG.2.2.27729.95843, https://www.researchgate.net/publication/380397398.

16. CMS collaboration, "A profile likelihood approach to measure the top quark mass in the lepton + jets channel," April 2022,

https://home.cern/news/news/physics/cms-measures-mass-top-quark-unparalleled-accuracy.

17 M. Khodaverdian, "Accuracy and Precision of the Z Boson Mass Measurement with the ATLAS Detector," May 27, 2019,

https://indico.cern.ch/event/813935/contributions/3557802/attachments/1919010/3174010/Gymnasieprojekt_Mariam_Khodaverdian_2019.pdf.

18. CDF Collaboration, "High-precision measurement of the W boson mass with the CDF II detector," April 2022, DOI: 10.1126/science.abk1781, https://www.science.org/doi/10.1126/science.abk1781.

19. W. M. Yao et al. (Particle Data Group), J. Phys. G33, 1 (2006) and 2007 partial update for edition 2008 (URL: http://pdg.lbl.gov).

20. R.L. Workman et al. (Particle Data Group), Prog. Theor. Exp. Phys. 2022, 083C01 (2022) and 2023 update.

21. R. Bayes, "Experimental Constraints on Left-Right Symmetric Models from Muon Decay," Physical Review Letters, January 28, 2011, https://www.researchgate.net/profile/Vladimir-Selivanov/publication/50397381.

22. Heyde, K. (2004), "Beta-decay: the weak interaction at work," *Basic Ideas and Concepts in Nuclear Physics: An Introductory Approach* (third ed.), Taylor & Francis, DOI:10.1201/978142005494.

23. R. Bayes et.al., "Experimental Constraints on Left-Right Symmetric Models from Muon Decay," Physical Review Letters, January 28, 2011, DOI: 10.1103/PhysRevLett.106.041804.

24. DT Froedge, "The Dirac Equation and the two Photon Model of the Electron revised," April 2021, , DOI: 10.13140/RG.2.2.19095.70564, www.researchgate.net/publication/350922403.

25. The KATRIN Collaboration, "Direct neutrino-mass measurement with sub-electronvolt sensitivity," *Nat. Phys.* 18, 160–166 (2022), https://doi.org/10.1038/s41567-021-01463-1.

26. DT Froedge, "Relativistic Time Dilation Illustrated in Delta-c Mechanics, The Twin Paradox is an Illusion," December 2023, DOI: 10.13140/RG.2.2.15584.46085, https://www.researchgate.net/publication.

27. DT. Froedge, "Neutrino Binding Between Nuclear Particles in Delta-c Mechanics Addendum to: Nuclear Particle Structure in Delta-C Mechanics," May 2024, DOI:10.13140/RG.2.2.27729.95843, https://www.researchgate.net/publication/380397398.

Appendix I

The Physical Constants and Particle Mass Values
from Codata and the Particle Data Group Used in This Paper

Particle	Particle Mass (gms)	Particle Mass (Electron)	Compton Radius (cm)
Electron	9.10938370150000E-28	1.0000000000000	3.86159267943989E-11
Muon	1.88353162700000E-25	206.7682829838260	1.86759430591295E-13
Proton	1.67262192369000E-24	1836.1526734400000	2.10308910326354E-14
Tauon	3.16754000000000E-24	3477.1882507793300	1.11055036503721E-14
Neutal Pion	2.40618001661378E-25	264.1430085130310	1.46193257250244E-13
Chgd Pion	2.48806333627759E-25	273.1319063734130	1.41381969273136E-13
Neutron	1.67492749804000E-24	1838.6836617324600	2.10019415509539E-14
Wboson	1.43344876621262E-22	157359.577023480	2.45399279311973E-16
Zboson	1.61700283014969E-22	177509.575086120	2.17542781991699E-16

Physical Constants

λ_{PL}	Planck Radius	1.61640095996445E-33 (cm)
α	Fine Structure Constant	7.29735253594845E-03
$\hbar$	Planck Constant	1.05457181760000E-27 (cgs)
c_0	Velocity of light	2.99792458000000E+10 (cm/sec)
ν_e	Electron Freq.	1.23558996386000E+20 (hz)
m_e	Electron Mass eV	5.10998902000000E+05 (ev)
G	Gravitational Constant	6.67550533180000E-08 $cm^3/gm\ sec^2$

Composit Constnts

$\overline{\nu}_e$	Electron cycle number	1.23558996386000E+20 Eq 4
E_0	Nuc Atom Ground State	2.67493983646500E-05 Eq 5
$\mathfrak{R}_0$	ECR Radius	2.82447977709503E-13 (cm)Eq 3
g_A	Anom Gyromagnetic Ratio	1.00115965218073 eq 11
η	Nuclear Anomalous Spin	1.00060014721177 Eq 12

1.00060127388753 Will this cause problem

83

Appendix II

Eddington

From Sir Arthur Eddington's book, *Relativity Theory of Photons and Electrons*, 1936, on wavefunctions.

Multiplication of Probabilities

It is of course entirely **opposed to our habit of thought to regard a system as the product of its parts rather than the sum of its parts.** Therefore **we have a long way to travel before we can connect the combination of systems by multiplications** with our ordinary outlook.

Admitting that multiplication is the primary operation in the theory (QM), the prevalence of exponentials and the formula which we have developed indicates that the way in which the subsidiary operations of addition arise. We see that in general the **additive quantities of physics must occur in** exponentials.

Appendix III

Deficit Energy

Through observation of the vertex function that makes up the mass of particles, it is apparent that, except for QED effects, the core of all is the same as the electron. It is thus presumed that vertex Kernels are electrons occupying different states.

When an electron falls into the ground state of hydrogen, the energy of that state (12.6 ev) is radiated away. The mass of the electron in that state is reduced by:

$$m' = m_e \left(\alpha^2 / 2 \right) \qquad (49)$$

The particle created, Δm, a photon that is radiated away, is in fundamental terms:

$$\frac{\Delta m}{m_e} = \left(\frac{1}{2} \left[\frac{\mathfrak{R}_0}{\lambda_e n_R} \right]^2 \right) \tag{50}$$

The particles are attractive and, on locking into a state, radiate the energy away as a photon, leaving the reduced electron m' bound in the state.

$$\frac{\mathfrak{R}_0}{\lambda_e n_R} = \frac{L}{\hbar\, n_R} \tag{51}$$

Nuclear particles operate somewhat the same; this can be demonstrated with the vertex function of the Muon. From Eq. (21), with some rearranging, the vertex function is:

$$\frac{m_\mu}{m_e} = \frac{1}{\dfrac{1}{\sqrt{2}} \left(\dfrac{\mathfrak{R}_0}{\lambda_e} 2^4 \right) \left(\dfrac{\mathfrak{R}_0}{\lambda_e} 2^4 \right)} = 206.768 \text{ electrons} \tag{52}$$

When the two vertexes come together, the energy of the binding creates the muon, but unlike the atomic particles in Eq. (50), there is nothing from which to draw the energy. The particles are repulsive, and the energy to create the particles has to be injected by inertia or pressure. Once the particles are bound however, the deficit of angular momentum holds the particles together, and the energy of the binding is radiated away.

The particles inertial mass exists as a deficit of energy. The Einstein mass energy relation would then be mass created by a deficit of energy:

$$mc_0^2 = -\varepsilon \tag{53}$$

Energy has not been lost but escaped to the universe. For a particle to be extracted from a nuclear state requires the energy equivalent to its mass be restored. Once restored, it can eject a free electron.

This relation provides the energy balance in the universe between the inertial mass and electromagnetic energy.

Appendix IV

Illustrated Supplement

A graphical illustration of the significant aspects of the vertex functions is presented here to better give an understanding of the process.

Noting that the Eddington view of the mass as a product of mass functions, in this case, the vertex kernels are the mass components of the vertex functions.

$$\left(\frac{m}{m_e}\right) = \left(\frac{m}{m_e}\right)\left(\frac{m}{m_e}\right)\left(\frac{m}{m_e}\right)\cdots \tag{1.54}$$

We can illustrate the relation between the atomic and nuclear vertex functions by comparing the atomic vertex function, the Muon vertex function, and the function of the lowest energy nuclear particle. From the text:

$$\frac{\varepsilon_R}{\varepsilon_e} = \frac{1}{2}\left(\frac{\alpha}{n_R}\right)^2 = \frac{1}{2}\left(\frac{\mathfrak{R}_0}{\lambdabar_e g_A^2 n_R}\right)^2 \quad \sim \rightarrow \quad \left(\frac{L_e}{\hbar\, n_R}\right)^2$$

$$\left(\frac{m_\mu}{m_e}\right) = 1 \, / \left[\frac{1}{\sqrt{2}}\left(\frac{\lambdabar_{PL}\, \bar{v}_e\, 2^4}{\lambdabar_e \eta}\right)\left(\frac{\lambdabar_{PL}\, \bar{v}_e 2^4}{\lambdabar_e \eta}\right)\right] \quad \sim \rightarrow \quad \left(\frac{\hbar}{L_e n_N}\right)^2$$

$$(1.55)$$

The ratio of the electron creation radius to the electron Compton radius is actually the ratio of its angular momentum to the Planck constant, $\mathfrak{R}_0 / \lambdabar = \mathfrak{R}_0 m_e c_0 / \hbar = L_e / \hbar$. This is the ratio of the angular momentum of the electron orbiting at the electron creation radius to the Planck action $\hbar$.

Atomic (Rydberg $n_R = 1$) state
$$\frac{\varepsilon_R}{\varepsilon_e} \rightarrow \left(\frac{\mathfrak{R}_0}{\lambdabar_e\, n_R}\right)^2 \rightarrow \left(\frac{1}{193.3 n_R}\right)^2 = 13.6\,\text{ev}$$

Nucler (Muon $n_N = 16$) states
$$\frac{m_\mu}{m_e} \rightarrow \left(\frac{\lambdabar_e}{\mathfrak{R}_0 n_N}\right)^2 \rightarrow \sqrt{2}\left(\frac{193.3}{n_N}\right)^2 = 207$$

(1.56)

Note the mass energy of the atomic Rydberg states and the mass of the nuclear state are virtually reciprocals.

Atomic states have integral values of $n\hbar$, whereas nuclear states have fractional values $\hbar/n$. The illustration of the position of the atomic and nuclear states are illustrated in Fig. 1A.

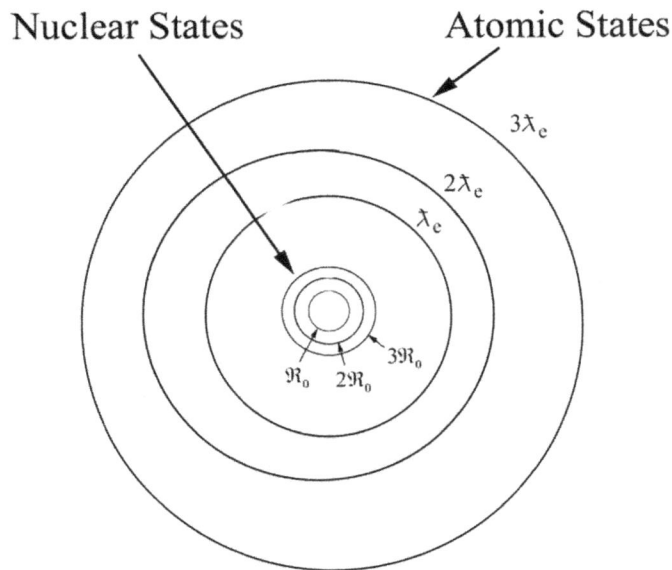

Nuclear States   Atomic States

$3\lambdabar_e$
$2\lambdabar_e$
$\lambdabar_e$
$3\mathfrak{R}_0$
$\mathfrak{R}_0$ $2\mathfrak{R}_0$

Relative locations of Atomic and Nuclear states, showing the consilience between the atomic (photon) and nuclear inertial mass (particle).

Fig. 1A

87

The similarity of the neutron and charged pion are illustrated in Fig. 1A and 2A.

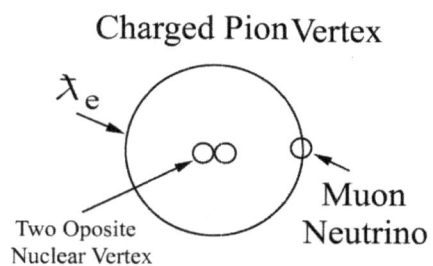

Charged Pion Vertex

λ_e

Two Oposite
Nuclear Vertex

Muon
Neutrino

Neutron Vertex

λ_e

Proton
Vertex

Electron
Neutrino

Fig. 2A

Relative locations of states
in the charged pion

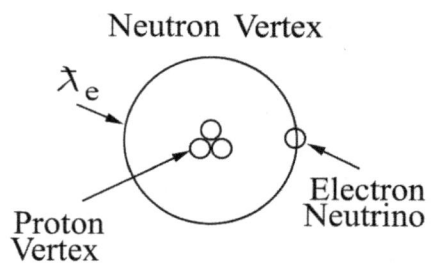

Fig. 3A

Relative locations of states
in the neutron

Structure of Elementary Nuclear Particles in Delta-c Mechanics

D.T. Froedge

V111324

Introduction

In previous research, the mass values of particles within the atomic nucleus have been presented. However, this paper aims to delve deeper by exploring the internal structure of these particles and elucidating why elementary mass particles exhibit specific, quantifiable values. The foundational studies, "Nuclear Particle Structure in Delta-C Mechanics" and "Neutrino Binding Between Nuclear Particles in Delta-C Mechanics" [13, 14], laid the groundwork for this ongoing investigation, providing the initial direction for understanding nuclear particle interactions. [2]

A common presumption in the field has been that the energy levels of particles are determined solely by potential energy, much like the quantized energy levels of electrons in atoms. According to this view, nuclear particles occupy specific potential wells that correspond to quantum mechanical conditions, where each particle is assigned a distinct place defined by its potential. However, this assumption oversimplifies the complexity of mass particle formation and fails to capture the true underlying mechanisms.

In contrast, this paper demonstrates that the mass values of elementary particles are governed not merely by potential levels but by intricate mechanical interactions and standing wave dynamics between the particles themselves. Through these mechanical interactions, particles achieve specific mass values that are not attributed to potential energy, but a deficit of energy.

The binding of two opposite ± particles is equivalent to the escape energy. Thus, as two free particles bind together, the mass is reduced. When the binding is closer than the radius of the electron, more energy is radiated away than their free particle mass and, on acquiring a state, exists as a deficit in energy. The mass of a nuclear particle is then:

$$m = -E/c^2 \qquad (1)$$

This is the deficit energy required to remove a particle from a nucleus, and perhaps explains the balance of energy in the universe between inertial mass energy and electromagnetic energy.

By examining the mechanical and wave-like interactions of the path density flows at the subatomic level, it is uncovered that the generation of elementary particle mass arises from a more nuanced process than previously assumed. It is the collective and dynamic interplay of particles—including their relation with electrons—that contributes to the mass values observed in nuclear particles, rather than a simple potential energy framework.

Charge

The concept of electric charge has long been understood as a fundamental property of matter that governs the electrical interactions between particles. Traditionally, it is believed that this property, often referred to simply as "charge," plays a critical role in the way particles attract or repel each other. In this framework, charged particles create electric fields that influence their surroundings, leading to observable forces. Electric repulsion is not present inside the Compton radius of a composite particle because the external probability flow is symmetric and has no differential effect on internal parties. Opposite particle pairs existing inside the Compton radius (electrons-positrons) cannot annihilate, since their energy is less than their free energy.

Building on this foundation, the theory of Delta-c Mechanics introduces a novel perspective by positing that the interaction between particles is fundamentally linked to the flow density and direction of photons along various action paths. The probability distribution of photon flow density interacts with the flow density of other particles, effectively altering the "index of refraction" of the medium through which photon probability move. This change in the index of refraction significantly changes the mechanism of how particles interact with one another, providing an alternate understanding of electrical interactions. In this view, the dynamics of photon flow not only mediate particle interactions but also transform our comprehension of charge to physical phenomena.

Preliminaries

Photons and Particles

Delta-c Mechanics postulates the photon is a rotating Planck particle having a radius of λ_{PL} (10E-33 cm), and a probability flow direction rotating at the Compton frequency, having energy of $\varepsilon = \hbar\omega$. If the photon is not in rotation, it has no energy but still has an interaction with the probability flow of other photons, changing the flow direction, thus having an effect on the index of refraction of other action photons and particle direction.

The primary rest mass particle is the electron, formed by the self-binding of two photons, which having sufficient probability flow to direct by the change in the index of the other into an orbit (see [5] for details). The change in the velocity of light near a rotating particle due to the probability of its photon location is:

$$\frac{\Delta c}{c_0} = \left(\frac{\lambda_{PL} \bar{v}_e}{r} \right) \quad (2)$$

This is the change in the velocity of light at a distance r from the rotation of a Planck particle at the repetition rate of the electron Compton frequency.

The product of two particles gives the mutual reduction of the change in c, which reduces the mutual mass energy by $\Delta\varepsilon$. The structure of this is discussed in. [12]

$$\frac{\Delta c}{c_0} = \frac{\Delta\varepsilon}{\varepsilon_0} = \frac{1}{2}\left(\frac{\sqrt{2}\lambda_{PL}\bar{v}_e}{r}\right)\left(\frac{\sqrt{2}\lambda_{PL}\bar{v}_e}{r}\right) = \left(\frac{\lambda_{PL}\bar{v}_e}{r}\right)\left(\frac{\lambda_{PL}\bar{v}_e}{r}\right)$$

(3)

This relation shows the change in the velocity of light at one particle at a distance r from the, and giving the interaction energy $\Delta\varepsilon$ as a result of the particle biding. $\Delta\varepsilon$ is the deficit in energy created by the binding, and is radiated away when the particles have a an integral $\hbar$ solution $n\hbar = mcr$. The numerators in Eq. (2) are the Electron Creation Radius, which is the radius of the two photons forming the first stable elementary inertial mass particle, the electron:

$$\mathfrak{R}_0 = \sqrt{2}\lambda_{PL}\bar{v}_e$$

(4)

The Planck particle radius, λ_{PL}, is the radius of the bare photon and defined in elementary terms as $\lambda_{PL} = \sqrt{G\hbar/c^3}$.

$\bar{v}$ is the ratio of the velocity light travels in free space to the relative velocity of the photons in the electron Compton orbit. It is "unitless," but its magnitude is equivalent to the electron frequency.

$$\bar{v}_e v_e = c_0 = v_e \lambda_e \rightarrow \bar{v}_e = \frac{c_0}{v_e}$$

(5)

The relations above are developed in [5].

92

The Ground State

For atomic as well as nuclear particles, Eq. (2), the binding energy ratio of two particles has solutions at integral values of the action quantum $\hbar$. The value of the binding of two electrons forms the ground state, E_0, of the atomic-nuclear particles. The nuclear ground state, E_0, is a unitless ratio of the binding energy of two $\pm$ electrons to the mass of an electron.

$$E_0 = \frac{1}{2}\left(\frac{\mathfrak{R}_0}{\lambda_e}\right)\left(\frac{\mathfrak{R}_0}{\lambda_e}\right) = \left(\frac{\lambda_{PL}\bar{v}_e}{\lambda_e}\right)\left(\frac{\lambda_{PL}\bar{v}_e}{\lambda_e}\right) = \frac{Gm_e}{\lambda_e}\left(\bar{v}_e\right)^2 \quad (6)$$

The fundamental value of the electron creation radius, $\mathfrak{R}_0$, is shown in Eq. (3), and has been expressed in more fundamental terms in Eq. (6). Note that the value E_0 is a creation of a particle with the energy extracted from the mass of the bound particles. When bound, this energy is radiated away leaving the bound pair having less mass than the free particles. The Ground state of nuclear particles is the solution when the radius corresponds to the free electron's Compton radius. This state marks the ground state of both atomic and nuclear particles.

The base state of any particle, like the electron, as defined in Eq. (6), is the binding of a positive and negative particle. If designated particle x, then the nuclear base state and the binding state of the x particle and E_0 is:

$$E_0 = \frac{1}{2}\left(\frac{\mathfrak{R}_0}{\lambda_e}\right)^2 \qquad\qquad E_x = \frac{1}{2}\left(\frac{\mathfrak{R}_0}{\lambda_x}\right)^2 \quad (7)$$

Escape and Deficit Energy

The integral of the potential energy from λ_e to infinity is the escape energy of an electron from the nuclear ground state.

$$\frac{\varepsilon_0}{m_e c_0^2} = \int_{r=\lambda_e}^{\infty} \left[\frac{1}{2} \left(\frac{\mathfrak{R}_0}{r} \right)^2 \right] = \frac{1}{2} \left(\frac{\mathfrak{R}_0}{\lambda_e} \right) = 0.0036571 \text{ electrons} \quad (8)$$

When the particle solution or Compton radius is less than $\lambda_X < \mathfrak{R}_0 / 2$ and the particle is bound in that position, the escape energy, equal to the binding energy, is radiated away. The particle mass thus exists as a "deficit energy."

The Rydberg energy ground state of atomic particles is not at the radius λ_e, but because of the QED-induced delay in the completion of the Compton radius, this makes the radius greater by the factor of the Anomalous Gyromagnetic Ratio, g_A. The Rydberg ground state $n_K = 1$ is at a radius of $r = \lambda_e g_A^2$.

If this radius is set as the solution to Eq. (2), the energy levels are the ionization energy of two opposite $\pm$ electrons, or twice the ionization energy of positronium.

$$\frac{E_0}{g_A^4} = \frac{1}{2} \left(\frac{\mathfrak{R}_0}{\lambda_e g_A^2} \right) \left(\frac{\mathfrak{R}_0}{\lambda_e g_A^2} \right) = \frac{1}{2} \alpha^2 \rightarrow 13.606 \text{ eV} / m_e \quad (9)$$

Inserting $\mathfrak{R}_0$ from Eq. (3) and the Rydberg integers, the Rydberg series gives the relation in fundamental terms as:

$$\frac{E_0}{g_A^4 n^2} = \frac{1}{2} \left(\frac{\lambda_{PL} \bar{v}_e}{\lambda_e n_{R1} g_A^2} \right) \left(\frac{\lambda_{PL} \bar{v}_\theta}{\lambda_e n_{R1} g_A^2} \right) = \frac{1}{2} \frac{\alpha^2}{n_{R1} n_{R1}} \quad (10)$$

This is the atomic Rydberg levels of the atom. The Rydberg integers n_R are multiples of Planck's constant in the Compton radius of the electron. Note that the product of the Rydberg integers n_R must also be an integer, $n_{R1} n_{R1} = n_R$, and has a physical cause in the binding of nuclear particles (see Binding Interaction Mechanism section, below).

The expressions in parentheses in Eq. (8) represents what is called the "Kernels of particle creation," K, forming the binding mechanism for both atomic and nuclear particles.

This function, distinguished by an integer value of the Rydberg integer, is central to both atomic and nuclear particles. While hydrogen atomic states exhibit only two particles in the binding, nuclear particles demonstrate multiple bindings of particles with the same Kernel. The multiplicity defined by n_K in Eq. (8) constitutes the structure of atomic particles.

Particle Masses

In Eq. (7), it can be observed that the energy created by the binding is extracted from the rest mass of the bound electrons, thus the mass of the orbiting particles have decreased. On acquiring a Rydberg state with $n_R \hbar$, the binding energy is radiated away and m', the bound particle mass, is as shown here:

$$m' = m_e - \frac{1}{2}\left(\frac{\alpha}{n_K}\right)^2 m_e \quad \rightarrow \text{Photon Radiated Energy} = \frac{1}{2}\left(\frac{\alpha}{n_K}\right)^2 = \frac{m_e - m'}{m_e}$$

$$(11)$$

In Eq. (7), when the radius of the particle solution λ is greater than $\mathfrak{R}_0$ or $r = \lambda_X > \mathfrak{R}_0$, the particle is bound in the Rydberg state n_R and the energy radiated away is less than the mass energy of the free electron (Atomic).

When the state energy has a solution with a Compton radius that is less than $\mathfrak{R}_0$, that is $r < \mathfrak{R}_0$, the energy radiated away is greater than the electron mass (Nuclear):

$$\text{Radiated Energy} \qquad \frac{m_e - m'}{m_e} \quad \rightarrow \quad -\frac{m' - m_e}{m_e} \qquad (12)$$

$$\text{Atomic} \qquad\qquad \text{Nuclear}$$

In Eq. (12), $m' < m_e$, the left relation is the energy in the atomic system in which the particles are attractive, and on acquiring the state, the energy radiated away leaves a particle with less energy than the free electron.

The second is the nuclear particle in which the particle mass is greater than the electron. On acquiring the state, the energy is radiated away, leaving the mass of the particle as a creation of a deficit in energy.

It is important to emphasize that this process does not imply negative energy. Instead, the radiated energy is released into the universe as electromagnetic radiation. However, it does imply positive inertial mass is created by an energy deficit. Since the inertial mass of all particles is positive, we can assert that the Einstein Energy Relation for nuclear particles should be:

$$m = -\varepsilon / c_0^2 \qquad (13)$$

This goes to the creation of atomic particles as being put together in an environment having a very low velocity of light and now cannot escape due to a deficiency of energy.

Differential Atomic Levels

For familiarity, including the Bohr-Sommerfeld atomic states can be expressed as:

$$\frac{E_{nl}}{\left(m_e c_0^2\right) g_A^4} = \frac{1}{2}\left(\frac{\alpha}{n_R}\right)^2\left[1+\left(\frac{\alpha}{n}\right)^2\left(\frac{n}{l+1/2}-\frac{3}{4}\right)\right]+O\left[\alpha^6\right] \qquad (14)$$

Note that the Anomalous Gyromagnetic factor g_A^4, appears as the base value and doesn't show in the radiation differentials.

Structure of Nuclear Particles

The base state of all particles as discussed is E_0, that is the binding energy of two free electrons Eq. (6). The radius λ_e of the Rydberg atomic ground levels is slightly larger than the free electron due to the QED-induced rotational delay of the Anomalous Gyromagnetic ratio g_A, this delay causes the Rydberg ground level to be at a larger radius of $\lambda_R = \lambda_e g_A^2$. This makes the ground energy of the Rydberg levels slightly less than the nuclear ground state energy, E_0. The exponent n of Eta. $g_A^{\pm n}$ depends on the number of mass particles bound with the electron.

There are two anomalous ratios that play a role in nuclear-atomic systems. They are the QED-determined loop integral delay for particles moving in a circle. The ratio has no effect on the free electron, but when it is bound in rotation, the Feynman action integral loops cause a delay that alters the electrons Compton radius. For atomic or Rydberg levels with radii above the electron, at $r = \lambda_e g_A^2$, the Anomalous Gyromagnetic ratio g_A^2 increases the electron Compton radius. Its value is:

$$g_A = 1.0011596521807 \tag{15}$$

The second is a similar factor that applies to particles with Compton radii less than $\mathfrak{R}_0$, is here designated as Eta. η. The value can be determined with precision by relations between the mass of several elementary particles and it is:

$$\eta = 1.00060014721177 \tag{16}$$

For nuclear particles, with Compton radii less than $\mathfrak{R}_0$, it is found that the QED effect on the radius is about the square root of the Anomalous Gyromagnetic ratio.

The Nuclear Anomalous ratio Eta. η represents the change in the Compton radius for a mass $\lambda'_x = \lambda_x \eta$ particle bound within the nucleus. This factor is consistently present in the mass of all nuclear states. As the value of the Anomalous Gyromagnetic ratio is very precise, it also is very precise. Eta. (η) is exact to at least twelve significant digits and uniform across all nuclear particles. The exponent of Eta. $\eta^{\pm n}$ also depends on the number of mass particles bound and the sign depends on relative spin.

Although the value of η has not yet been determined by QED methods, it should not be dismissed as arbitrary, as it is crucial to the precise mass of several interconnected known particles. This factor, like the Anomalous Gyromagnetic ratio, is an offset to the nuclear ground state. Originally, this factor was mistakenly presumed to be a neutrino (see Appendix I).

The Particle Kernels

From Eq. (8), the Kernel of atomic and nuclear particles, K, is identified as:

$$K_{(\text{Atomic})} = \left(\frac{\lambda_{\text{PL}} \ \overline{v}_e}{\lambda_e g_A^2 \ n_R} \right) \qquad K_{(\text{Nuclear})} = \left(\frac{\lambda_{\text{PL}} \ \overline{v}_e}{\lambda_e \eta \ n_R} \right) \qquad (17)$$

Atomic Electrons   **Nuclear Quarks**

The product of these relations establishes particle masses. Note that the Rydberg integer n_R is a multiple of the action quantum ($n_R \lambda = n_R \hbar / mc$) and plays the same role in nuclear particles as in atomic states. The kernels are the same except for the QED effects. The value of η in the kernel can be $\eta^{\pm n}$, indicating an increase or decrease in the electron radius, presumably the result of the relative spin between different particles.

All nuclear particles exhibit similar binding mechanisms. While the atomic system showcases the binding of two opposite electron-positrons, other particles have base states also that are related to E_0.

The forms illustrating the interacting binding of two particles in Eq. (5) are; first, in the atomic form $\lambda_e > \mathfrak{R}_0$ and; second, applies to particles with mass greater than the electron $\lambda_e < \mathfrak{R}_0$ in the nuclear form.

The ratio of the binding energy of the electrons to the binding energy of an arbitrary particle x yields the mass of the x particle in electrons.

$$\frac{E_0}{E_x} = \frac{\frac{1}{2}\left(\frac{\mathfrak{R}_0}{\lambda_x}\right)^2}{\frac{1}{2}\left(\frac{\mathfrak{R}_0}{\lambda_e}\right)^2} \quad \rightarrow \quad \frac{1}{2}\left(\frac{E_x}{E_0}\right) = \left(\frac{m_x}{m_e}\frac{m_x}{m_e}\right) \quad \rightarrow \quad \sqrt{2\frac{E_x}{E_0}} = \frac{m_x}{m_e}$$

(18)

E_0 is the ground state of the nuclear-atomic system. The ratio here is the binding energy of positive & negative x particle energy relative to the ground state energy. The last term in Eq. (14) is the expression solved for the mass of the single x particle.

Binding Kernels

(redefined in later version)

As atomic bindings create a state particle, namely photons, from the binding of two electrons, the binding of two or more sub-particles defined by the Kernels in Eq. (13) can create state particles.

Atomic

$$E_x = E_0\left[(K_1)(K_2)\right]$$

$$\frac{m_x}{m_e} = \sqrt{\left[(K_1)(K_2)\right]}$$

Nuclear

$$E_x = \frac{E_0}{(K_1)(K_2)(K_3)}$$

$$\frac{m_x}{m_e} = 1/\sqrt{(K_1)(K_2)(K_3)}$$

(19)

The difference is that in the atomic case, the particle can radiate away, reducing the mass of the bound particle, but for a nuclear particle, the energy radiated away exceeds the free particle rest mass and thus the bound particles are trapped as deficits in energy. All sub-particles have the same kernel as electrons with the difference being the QED factors g_A and η.

Including the electron in the Rydberg calculations shows binding radiated energy.

Atomic

$$\left[\begin{array}{l} \Delta E_X = E_0 \left[1 - \left(\dfrac{\sqrt{2}\lambda_{PL}}{\lambda_e g_A^2} \dfrac{\bar{v}_e}{n_{K1}} \right) \left(\dfrac{\sqrt{2}\lambda_{PL}}{\lambda_e g_A^2} \dfrac{\bar{v}_e}{n_{K2}} \right) \right] \\[2ex] \rightarrow \dfrac{m_e^2 - m_x^2}{m_e^2} = \left(\dfrac{\sqrt{2}\lambda_{PL}}{\lambda_e g_A^2} \dfrac{\bar{v}_e}{n_{K1}} \right) \left(\dfrac{\sqrt{2}\lambda_{PL}}{\lambda_e g_A^2} \dfrac{\bar{v}_e}{n_{K2}} \right) \\[2ex] \rightarrow \dfrac{(m_e - m_x)}{m_e} = \dfrac{1}{2} \left(\dfrac{\sqrt{2}\lambda_{PL}}{\lambda_e g_A^2} \dfrac{\bar{v}_e}{n_{K1}} \right) \left(\dfrac{\sqrt{2}\lambda_{PL}}{\lambda_e g_A^2} \dfrac{\bar{v}_e}{n_{K2}} \right) \\[2ex] \rightarrow \dfrac{\Delta c}{c_0} = \dfrac{(m_e - m_x)}{m_e g_A^4} = \dfrac{1}{2} \left(\dfrac{\alpha^2}{n_K} \right) \qquad \begin{array}{l}\text{Radiated Energy}\\ \text{See Equation (11)}\end{array} \end{array}\right] \tag{20}$$

The above particle created is the photon of the radiated energy.

Nuclear

$$\left[\begin{array}{c} E_X = E_0 \left(\dfrac{\sqrt{2}\lambda_{PL}\eta}{\lambda_e} \dfrac{\bar{v}_e}{n_{K1}} \right)^{n1} \left(\dfrac{\sqrt{2}\lambda_{PL}\eta}{\lambda_e} \dfrac{\bar{v}_e}{n_{K2}} \right)^{n2} \left(\dfrac{\sqrt{2}\lambda_{PL}\eta}{\lambda_e} \dfrac{\bar{v}_e}{n_{K2}} \right)^{n3} \\[3ex] \dfrac{E_X}{E_0} = \left(\dfrac{m_x}{m_e} \right)^2 \\[3ex] \dfrac{m_x}{m_e} = 1 \left/ \sqrt{\left(\dfrac{\sqrt{2}\lambda_{PL}\eta}{\lambda_e} \dfrac{\bar{v}_e}{n_{K1}} \right)^{n1} \left(\dfrac{\sqrt{2}\lambda_{PL}\eta}{\lambda_e} \dfrac{\bar{v}_e}{n_{K2}} \right)^{n2} \left(\dfrac{\sqrt{2}\lambda_{PL}\eta}{\lambda_e} \dfrac{\bar{v}_e}{n_{K2}} \right)^{n3}} \right. \end{array}\right] \tag{21}$$

The nuclear mass created is equal to the energy radiated away on binding at an integral value of $\hbar$.

The simplified result of the internal mass of the particles could be stated as:

$$\frac{1}{2} \left(\frac{m_e}{m_x} \right)^2 = \left(\frac{m_e}{m_1} \right)^2 \left(\frac{m_e}{m_2} \right)^2 \left(\frac{m_e}{m_3} \right)^2 \tag{22}$$

The mass of m_x is not the sum of the particle mass but the product.

Binding Interaction Mechanism

Particles are bound to each other at the Compton radius in a plane and, as they spin, the interaction consists of simultaneous rotations intermeshing with another. For a stable solution representing an existing particle, this interaction manifests as a standing wave and continuity of probability flow between the two particles. For this to hold true, the number of electron cycles of one particle must be an integral multiple of the other; otherwise, a complete cycle cannot occur, and probability flow would not be conserved. This means the product of the Rydberg integer n_K of one multiplied by the other must also be an integer.

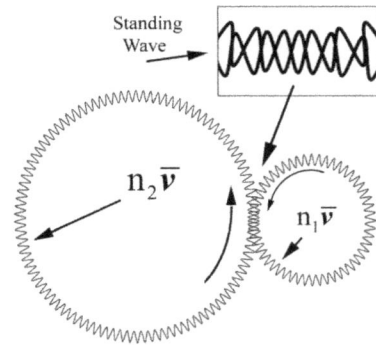

Fig. 1: The rotation connection of two bound particles

The interaction point is illustrated in Fig. 1, which depicts the continuous standing wave pattern at the contact point between the two particles. For the interacting particles, there has to be a continuous standing wave, thus the product of the Rydberg integer for two interacting particles also has to be an integer.

$$\frac{\bar{v}_e}{n_{K1}} \times \frac{\bar{v}_e}{n_{K2}} = \frac{\bar{v}_e}{n} \quad \text{or} \quad \left(\bar{v}_e n_{K1}\right) \times \left(\bar{v}_e n_{K2}\right) = \bar{v}n_e \qquad (23)$$

The value of n can be $n^{\pm 1}$, and there can possibly be many combinations specifying infinity of particles, but the interaction of particles having irrational products can't work. In some instances, two particles in a group can individually bind with another particle, but not with each other.

Calculated Values

The makeup of the following particles and calculations are shown below. Codata and experimental values determined by the particle data groups are included in Appendix IV. Those illustrated are:

Proton, Tauon, Muon, Charged Pion, Z Boson, W Boson, Top Quark

Summary

	Theory		Particle Mass Experimental, Codata, & PDG	
	Electrons	GeV	Electrons	
Proton	1836.15267344		1836.15267344	
Neutron	1838.68366173		1838.68366173	
Tauon	3477.18825		3477.135 +/- .136	
Muon	206.76743435		206.76828298	
Pion+/-	273.131906		273.1321133	
Z Boson	177190.369	90.544	90.04 - 90.76	GeV
W Boson	157359.58	80.410	80.433 +/-.0009	GeV
Top Quark	336153.82	171.7225	171.77 +/- 0.38	Gev

Proton

To illustrate the preceding mechanics, the mass of the proton can be shown.

The binding energy for the Proton as illustrated in Eq. (21) is:

$$\text{proton} \quad \frac{1}{2}\left(\frac{\Re_0}{\lambda_P}\right)^2 = \frac{\left(\dfrac{\Re_0}{\lambda_e}\right)^2}{\left(\dfrac{\lambda_{PL}\eta^3\,\bar{v}}{\lambda_e}\,\dfrac{}{2^5}\right)\left(\dfrac{\lambda_{PL}\eta^3\,\bar{v}}{\lambda_e}\,\dfrac{}{1}\right)^{1/2}\left(\dfrac{\lambda_{PL}\eta^3\,\bar{v}}{\lambda_e}\,\dfrac{}{2}\right)^{1/2}} \tag{24}$$

$$\left(\frac{m_P}{m_e}\right) = 1\left/\sqrt{\left(\frac{\lambda_{PL}\eta^3\,\bar{v}}{\lambda_e}\,\frac{}{2^8}\right)\left(\frac{\lambda_{PL}\eta^3\,\bar{v}}{\lambda_e}\,\frac{}{1}\right)^{1/2}\left(\frac{\lambda_{PL}\eta^3\,\bar{v}}{\lambda_e}\,\frac{}{2}\right)^{1/2}}\right. \tag{25}$$

Note that the elements of this are the same as electrons in atomic Rydberg states, and there are three particles, two with root exponents. It is asserted that these three elements are bound Quarks.

The values of the Rydberg integers for the Proton n_R for the three Quarks in the proton are 2^8, 1, and 2. Note that the first and third particle cannot combine without an irrational n_R product, $\sqrt{2}$ in the mass.

Using the data in Appendix IV, the values of the proton and quarks can be calculated to the precise value of the proton.

Product	Mass Mass in Electrons	Name
0.000323831	111.14006301801	Quark 1
0.017995311	3.7272654691143	Quark 2
0.025449213	4.4324906153747	Quark 3
1.48304E-07	1836.15267344223	Proton
	Proton within experimental error bars	

103

The identified quarks here are not the same as the experimental values of the Quarks as determined in the Standard model. ($u = 4.501, d = 9.393$)

Neutron

The structure of the neutron is the same as the proton with the added sub-particle composed of an electron in orbit at a Rydberg level $r = \lambda_e g_A^3$, (1), and another part that can probably be identified as a neutrino (2). The calculated value is shown to be as precise as the proton.

The product of the binding of atomic and nuclear Kernels from Eq. (15) is:

$$(2)\downarrow$$

$$\left(\frac{\mathcal{R}_0}{\lambda_e}\right)^2 = \left(\frac{\mathcal{R}_0}{\lambda_n}\right)^2 \frac{\left(\frac{\lambda_{PL}\eta^3\overline{v}}{\lambda_e}\right)}{\left(\frac{\lambda_{PL}\overline{v}}{\lambda_e g_A^3}\right)}\left(\frac{\lambda_{PL}\eta^3\overline{v}}{\lambda_e 2^6}\right)\left(\frac{\lambda_{PL}\eta^3\overline{v}}{\lambda_e}\right)^{1/2}\left(\frac{\lambda_{PL}\eta^3\overline{v}}{\lambda_e 2}\right)^{1/2} \tag{26}$$

$$(1)\uparrow$$

$$\left(\frac{m_X}{m_e}\right) = \sqrt{\frac{\left(\frac{\lambda_{PL}\overline{v}}{\lambda_e g_A^3}\right)}{\left(\frac{\lambda_{PL}\eta^3\overline{v}}{\lambda_e}\right)\left(\frac{\lambda_{PL}\eta^3\overline{v}}{\lambda_e 2^6}\right)\left(\frac{\lambda_{PL}\eta^3\overline{v}}{\lambda_e}\right)^{1/2}\left(\frac{\lambda_{PL}\eta^3\overline{v}}{\lambda_e 2}\right)^{1/2}}} \tag{27}$$

This gives 1837.6836, and adding the electron into the mass gives:

Product	Mass Mass in Electrons	Error
Neutron	1838.68366173244	Within experimental error

Muon

Two Ks with $n_R = 2^4$ and two Ks with $n_R = 2^3$.

$$\frac{1}{2}\left(\frac{\lambda_\mu}{\lambda_e}\right)^2 = \left(\frac{\lambda_{PL}\,\bar{v}_e\,2^4}{\lambda_e\,\eta}\right)^2 \times \left(\frac{\lambda_{PL}\,\bar{v}_e\,2^3}{\lambda_e\,\eta}\right)^2 \tag{28}$$

$$\frac{m_\mu}{m_e} = 1 \,/\, \sqrt{\frac{1}{2}\left(\frac{\lambda_{PL}\,\bar{v}_e\,2^4}{\lambda_e\,\eta}\right)^4} \tag{29}$$

Mass

Product of Terms	In Electrons	Experimental
0.000026845864459	136.47286	
0.319056188741313	1.2518474	
0.019312078693610	5.08827401	
1.65414504620E-07	3477.18825	3477.228+/- 0.459
		Within experimental error bars

This error of one part in 250,000 is at this point not explained.

$$\left(\frac{m_W}{m_e}\right) = \left(\frac{\hbar}{S_e}\frac{2^5}{\eta^3}\right)\sqrt{\left(\frac{\hbar}{S_e}\frac{2^6}{\eta^3}\right)^{1/2}\left(\frac{\hbar}{S_e}\frac{1}{\eta^3}\right)^{1/4}\left(\frac{\hbar}{S_e}\frac{2}{\eta^3}\right)^{1/4}}$$

Tauon

$$\frac{1}{2}\left(\frac{\lambda_\tau}{\lambda_e}\right)^2 = \left(\frac{\lambda_{PL}}{\lambda_e}\frac{\bar{v}_e}{2}\frac{\eta^3}{1}\right)^2 \left(\frac{\lambda_{PL}}{\lambda_e}\frac{\bar{v}_e}{1}\frac{\eta^3}{1}\right)^{3/4} \left(\frac{\lambda_{PL}}{\lambda_e}\frac{\bar{v}_e}{1/2}\frac{\eta^3}{}\right)^{1/4}$$

(30)

$$\frac{m_\tau}{m_e} = 1 / \sqrt{\left(\frac{\lambda_{PL}}{\lambda_e}\frac{\bar{v}_e}{1}\frac{\eta^3}{}\right)^2 \left(\frac{\lambda_{PL}}{\lambda_e}\frac{\bar{v}_e\eta^3}{1}\right)^{1/4} \left(\frac{\lambda_{PL}}{\lambda_e}\frac{\bar{v}_e}{2}\frac{\eta^3}{}\right)^{3/4}}$$

(31)

Product of Terms	Mass In Electrons	Experimental
0.000026845864459	136.47286	
0.319056188741313	1.2518474	
0.019312078693610	5.08827401	
1.65414504620E-07	3477.18825	3477.228+/- 0.459
		Within experimental error bars

Charged Pion

The pion appears to be the binding of two opposite quarks with an electron that orbits at the Electron Rydberg, making it a two-quark atom. The structure in the form of Eq. (15), as illustrated in Eq. (20) is:

$$\frac{1}{2}\left(\frac{\mathfrak{R}_0}{\lambda_e}\right)^2 = \frac{1}{2}\left(\frac{\mathfrak{R}_0}{\lambda_{\pi+}}\right)^2 \frac{\left(\dfrac{\lambda_{PL}}{\lambda_e \eta^2}\dfrac{\overline{v}_e}{1}\right)^4}{\left(\dfrac{\lambda_{PL}}{\lambda_e g_A^2}\dfrac{\overline{v}_e}{1}\right)^2} \tag{32}$$

Or:

$$\frac{m_{\pi+}}{m_e} = \frac{\alpha}{\left(\dfrac{\lambda_{PL}\overline{v}}{\lambda_e \eta}\right)^2} \tag{33}$$

Calculated Mass		Particle Data Group
Electrons	MeV	MeV
273.1319063	139.570104	139.57039± 0.00017 [19][20]

Note that the calculated mass value is at the edge of the experimental error bars for the charged pion by the PDG [18][19] but within the scatter of the latest experimental data (see Appendix III).

Z Boson

In [13], it had been found that the relation for the mass of the Z boson in terms of the proton and the nuclear ground state E_0 was:

$$\frac{1}{2}\left(\frac{\lambda_Z}{\lambda_e}\right)^2 = 2\left(\frac{\lambda_P}{\lambda_e}\right)^2 E_0$$

By putting the values for the proton and E_0 from above, the particles in the Z boson are:

$$\frac{1}{2}\left(\frac{\lambda_Z}{\lambda_e}\right)^2 = 1/\left[\left(\frac{\lambda_{PL}\eta^3 \overline{v}_e}{\lambda_e \quad 2}\right)^3 \left(\frac{\lambda_{PL}\eta^3 \overline{v}_e}{\lambda_e \quad 4}\right)^{1/2} \left(\frac{\lambda_{PL}\eta^3 \overline{v}_e}{\lambda_e \quad 2}\right)^{1/2}\right] \quad (34)$$

$$\left(\frac{m_Z}{m_e}\right) = 1/\sqrt{\left(\frac{\lambda_{PL}\eta^3 \overline{v}_e}{\lambda_e \quad 2}\right)^3 \left[\left(\frac{\lambda_{PL}\eta^3 \overline{v}_e}{\lambda_e \quad 4}\right)^{1/2} \left(\frac{\lambda_{PL}\eta^3 \overline{v}_e}{\lambda_e \quad 2}\right)^{1/2}\right]} \quad (35)$$

	Calculated Mass	Experimental Mass
Electrons	GeV	GeV
177190.369	90.544	90.04 - 90.76

Ref. [17]

Decay Channel	Entries	Center	Error	Center deviance from Z_m^0(%)	Z-score	Width	Error
Muon	600071	90.76	(6.03E-03)	-0.47	70.70	5.71	(2.38E-02)
Electron	375216	90.04	(8.90E-03)	-1.26	128.96	6.57	(3.68E-02)

W Boson

The W boson was found in [14] and [15] to be:

$$\frac{1}{2}\frac{\lambda_w}{\lambda_e} = \frac{\lambda_\tau}{E_0}\left(\frac{\lambda_P}{\lambda_e}\right)^2$$

By putting the values for the proton, tauon, and E_0 from above, the W boson particle makeup is:

$$\frac{1}{2}\left(\frac{\lambda_w}{\lambda_e}\right)^2 = \left[\left(\frac{\lambda_{PL}\eta^3}{\lambda_e}\frac{\overline{v}_e}{2^6}\right)^2\left(\frac{\lambda_{PL}\eta^3}{\lambda_e}\frac{\overline{v}_e}{2}\right)^{1/2}\left(\frac{\lambda_{PL}\eta^3}{\lambda_e}\frac{\overline{v}_e}{1}\right)^{1/4}\left(\frac{\lambda_{PL}\eta^3}{\lambda_e}\frac{\overline{v}_e}{2}\right)^{1/4}\right]$$

(36)

$$\frac{m_w}{m_e} = \sqrt{\left(\frac{\hbar}{S_e}\frac{2^5}{\eta^2}\right)\left(\frac{\hbar}{S_e}\frac{2^6}{\eta^2}\right)^{1/2}\left(\frac{\hbar}{S_e}\frac{1}{\eta^2}\right)^{1/4}\left(\frac{\hbar}{S_e}\frac{2}{\eta^2}\right)^{1/4}} = \frac{1}{2^{1/4}}\left(\frac{\hbar}{S_e}\frac{2^5}{\eta^2}\right)\left(\frac{m_P}{m_e}\right)^{1/2}$$

(37)

Calculated Mass		Experimental Mass	
Electrons	GeV	GeV	
157359.58	80.410	80.433Gev +/-9Mev	[17]

Fermilab (CDF) Collaboration 80.433Gev +/-9Mev

Top Quark

The Top Quark was found in [16] and [15] to be: $\lambda_{TOP} / \lambda_e = E_0 / 9$. Putting this in the same form as Eq. (15), gives:

$$\frac{1}{2}\left(\frac{\lambda_{TOP}}{\lambda_e}\right)^2 = \left(\frac{\lambda_{PL}\,\eta}{\lambda_e}\,\frac{\overline{v}_e}{6}\right)\left(\frac{\lambda_{PL}\,\eta}{\lambda_e}\,\frac{\overline{v}_e}{3}\right)^3 \qquad (38)$$

The nuclear energy of this particle is a deficit energy calculated from Eq. (38) to be:

$$\frac{m_{TOP}}{m_e} = 1 / \left(\frac{\lambda_{PL}\,\eta}{\lambda_e}\,\frac{\overline{v}_e}{3}\right)^2 = 336052.6 \text{ Electrons} \qquad (39)$$

	Calculated Mass	Experimental Mass
Electrons	GeV	GeV
336052.68	171.722	171.77+/-0.38 Gev
		CMS collaboration [16]

Conclusion

Without any injected coupling constants, the mass numbers are exceptionally accurate. Of the calculated mass values presented, only the muon mass is out by an unexplained, one part in one-fourth of a million. It's difficult to believe these calculated values without arbitrary form factors, having accuracies of up to twelve significant digits, are not the result of physical causation.

Earlier papers have focused on the physical aspects of Delta-c Mechanics [1], [2]. This paper's intention is to extend the foundational understanding of particle mass by focusing on the internal structure and mechanical interactions of elementary particles. Unlike prior models that primarily attributed mass values to potential energy levels, our exploration into Delta-c Mechanics reveals that mass states are governed by intricate standing wave dynamics between particle interactions. Key to this process is the electron, which establishes the ground state for all particles. The insights derived from the interactions between photons and particles shows the binding mechanisms and provides a clearer understanding of mass formation at both atomic and nuclear levels.

The finding that nuclear particles inertial mass is created by a deficit of energy produces an understanding of the balance of energy in the universe between electromagnetic and inertial mass.

It was illustrated that a proton is actually the same as two positrons and an electron trapped as an energy deficit of 1833.15 electrons. In the hydrogen atom, the other electron bound at the atomic level completes the sub-particles group, as two complete electron-positron pairs, with 1833 electron equivalent energy, radiated away.

References

DT Froedge, *The Physics of Delta-c Mechanics*, ISBN-13: 979-8218347178 (Feb. 14, 2024), (Papers 3–14 can be found as chapters in this publication), https://www.amazon.com/Physics-Delta-C-Mechanics-Approach-Particle/dp/B0CVZ8CNYQ.

1. DT Froedge, "The Concepts and Principles of Delta-c Mechanics," August 2024, DOI: 10.13140/RG.2.2.31630.37445, https://www.researchgate.net/publication/382850589.

2. DT Froedge, Acceleration Gravitation and Origin of Centrifugal Force in Delta-c Mechanics, July 2024, DOI: 10.13140/RG.2.2.28944.42247, https://www.researchgate.net/publication/381915343.

3. DT Froedge, "The Connection between Electric Charge, Gravitation, and the Feynman Sum over All Histories View of Quantum Electrodynamics," April 2020, Conference: APS, April 18–21, 2020, Washington, DC, https://absuploads.aps.org/presentation.cfm?pid=18355, https://www.researchgate.net/publication/341310206.

4. DT Froedge, "A Quantum Theory Conjecture on the Origin of Gravitational and Electric Particle Interaction," December 2019, DOI: 10.13140/RG.2.2.29097.54884, https://www.researchgate.net/publication/337826826.

5. DT Froedge, "The Electron as a Composition of Two Vacuum Polarization Confined Photons," April 2021, DOI: 10.13140/RG.2.2.18971.18722, https://www.researchgate.net/publication/350740864.

6. DT Froedge, "The Gravitational Constant to Eleven Significant Digits," March 2020, DOI: 10.13140/RG.2.2.32159.38564, https://www.researchgate.net/publication/339943651.

7. DT Froedge, "The Fine Structure Constant from the Feynman Path Integrals," March 2021, DOI:10.13140/RG.2.2.12979.55846, https://www.researchgate.net/publication/350188862.

8. DT Froedge, "Vacuum Polarization, Gravitation, Charge, and the Speed of Light," Sept. 2021, DOI:10.13140/RG.2.2.15619.22569, https://www.researchgate.net/publication/354474157.

9. DT Froedge, "The Calculated value of the Fine Structure Constant from Fundamental Constants," September 2021, DOI:10.13140/RG.2.2.34349.41440.

11 "CODATA values of the fundamental physical constants," https://www.nist.gov/programs-projects/codata-values-fundamental-physical-constants.

12. DT Froedge, "Electron Mass and State Energy Levels Resulting from Photon-Photon Interaction," Conference: APS, April 2022, New York, https://www.researchgate.net/publication/359912763.

13. DT Froedge, "Nuclear Particle Structure in Delta-C Mechanics," November 2023, DOI: 10.13140/RG.2.2.35774.05447.

14. DT Froedge, "The Gravitational Constant to Eleven Significant Digits," March 2020, DOI: 10.13140/RG.2.2.32159.38564, https://www.researchgate.net/publication/339943651.

15. DT Froedge, "Neutrino Binding Between Nuclear Particles in Delta-c Mechanics," May 2024, DOI:10.13140/RG.2.2.27729.95843, https://www.researchgate.net/publication/380397398.

16. CMS collaboration, "A profile likelihood approach to measure the top quark mass in the lepton + jets channel," April 2022,

https://home.cern/news/news/physics/cms-measures-mass-top-quark-unparalleled-accuracy.

17 M. Khodaverdian, "Accuracy and Precision of the Z Boson Mass Measurement with the ATLAS Detector," May 27, 2019,

https://indico.cern.ch/event/813935/contributions/3557802/attachments/1919010/3174010/Gymnasieprojekt_Mariam_Khodaverdian_2019.pdf.

18. CDF Collaboration, "High-precision measurement of the W boson mass with the CDF II detector," April 2022, DOI: 10.1126/science.abk1781, https://www.science.org/doi/10.1126/science.abk1781.

19. W.-M. Yao et al. (Particle Data Group), J. Phys. G33, 1 (2006) and 2007 partial update for edition 2008 (URL: http://pdg.lbl.gov).

20. R.L. Workman et al. (Particle Data Group), Prog.Theor.Exp.Phys. 2022, 083C01 (2022) and 2023 update.

Appendix I

Neutrino Misidentification Discussion

In earlier papers, "Neutrino Binding Between Nuclear Particles in Delta-c Mechanics," and "Nuclear Particle Structure in Delta-C Mechanics" [13][14], It was asserted that the slight offsets in the mass were the inclusion of the values of particular neutrinos. Although the developed values were substantially accurate, the structural mechanism was incomplete. The presentation here is more complete and the revised values are included.

The neutrinos carry off the extra energy on the breakup of a nuclear particle, but the radiated energy is not obvious as a state value in the particle.

The difference in the ratio of the free and bound energy is totally due to the change in the energy difference provided by the QED effects of the anomalous g_A factor, and Etaη, on the binding, and although it is not a neutrino, it is somehow related to the neutrino energy.

Appendix II

The following is the values of the mass of the relation between the nuclear ground state and four elementary particle masses presented in [15] and delineated in this paper. The integral numbers in these exact relations between the mass and the nuclear ground state give a clue to important relations developed here.

$$E_0 = 2.67493983646500E\text{-}05$$

$$E_0 = \frac{1}{2}\left(\frac{\Re_0}{\lambda_e}\right)^2 \qquad E_0 = \frac{1}{\sqrt{2^{15}}}\left(\frac{\lambda_\mu \eta^2}{\lambda_e}\right) \qquad E_0 = 2\left(\frac{\lambda_{\pi+}g_A^2}{\lambda_e \eta^2}\right)^2$$

$$E_0 = \sqrt{2}\left(\frac{\lambda_\tau \eta^{4.5}}{\lambda_e}\right)^{4/3} \qquad\qquad E_0 = \sqrt{2^{13}}\left(\frac{\lambda_P \eta^3}{\lambda_e}\right)^2$$

Appendix III

Pion Mass Error Bars

The pion mass error shown from Eq. (33) is at the edge of the R.L. Workman et al. (Particle Data Group), 2022 error bars for the value for the Pion mass, and at the center of the error bars in the 1008 publication. It is not believed that the calculated value of the pion mass in this paper is in disagreement with the experimental value, [18], [19].

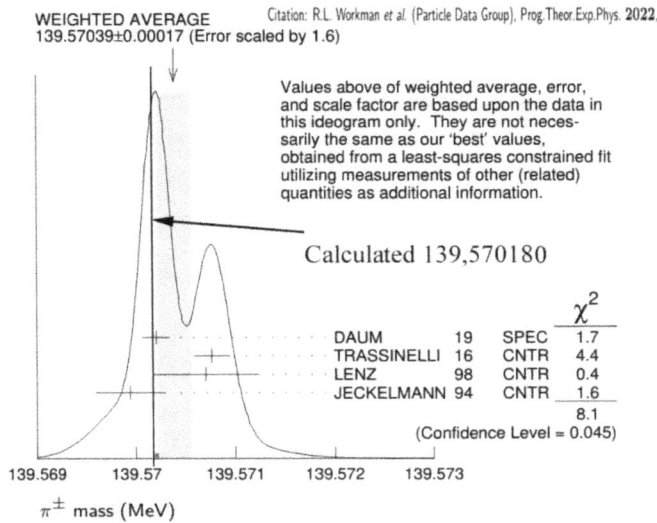

Comparison in GeV

Theory			PDG
139.57018		2008	139.57018± 0.00035
		2022	139.57039± 0.00017

Appendix IV

The Physical Constants and Particle Mass Values

Used in This Paper

Particle	Particle Mass (gms)	Particle Mass (Electron)	Compton Radius (cm)
Electron	9.10938370150000E-28	1.0000000000000	3.86159267943989E-11
Muon	1.88353162700000E-25	206.7682829838260	1.86759430591295E-13
Proton	1.67262192369000E-24	1836.1526734400000	2.10308910326354E-14
Tauon	3.16754000000000E-24	3477.1882507793300	1.11055036503721E-14
Neutal Pion	2.40618001661378E-25	264.1430085130310	1.46193257250244E-13
Chgd Pion	2.48806333627759E-25	273.1319063734130	1.41381969273136E-13
Neutron	1.67492749804000E-24	1838.6836617324600	2.10019415509539E-14
Wboson	1.43344876621262E-22	157359.577023480	2.45399279311973E-16
Zboson	1.61700283014969E-22	177509.575086120	2.17542781991699E-16

Physical Constants

λ_{PL}	Planck Radius	1.61640095996445E-33 (cm)
α	Fine Structure Constant	7.29735253594845E-03
$\hbar$	Planck Constant	1.05457181760000E-27 (cgs)
c_0	Velocity of light	2.99792458000000E+10 (cm/sec)
ν_e	Electron Freq.	1.23558996386000E+20 (hz)
$\bar{\nu}_e$	Electron cycle number	1.23558996386000E+20
m_e	Electron Mass eV	5.10998902000000E+05 (ev)
G	Gravitational Constant	6.67550533180000E-08 $cm^3 / gm\ sec^2$

E_0	Nuclear Ground State	2.67493983646500E-05
$\mathfrak{R}_0$	ECR Radius	2.82447977709503E-13 (cm)
g_A	Anom Gyromagnetic Ratio	1.00115965218073
η	Nuclear Anomalous Spin	1.00060014721177

The Concepts and Principles of Delta-c Mechanics

D.T. Froedge

V080324

Δc **Mechanics** is a reformulation of physics based on the Feynman formulation of Quantum Electrodynamics (QED). Feynman's QED posits that a particle moving from one point to another can be described as the sum of all possible action paths connecting these points. Essentially, the particle's trajectory is represented by integrating over all possible paths, each weighted by its action. In Feynman and wheeler's original view, the electron is the primary particle, and its interactions are mediated by the exchange of photons. [3]

Feynman's formulation of Quantum Electrodynamics (QED) yields extraordinarily precise predictions that have been confirmed with remarkable accuracy. In QED, particles move from one point to another through an infinite sum of all possible action paths. These paths, which represent possible trajectories for particles, sum to be the actual path of the particle. The action paths, which exist throughout space, are empty and have no effect on other particles except at the end points, where the particle is reconstituted as a solution to the Schrödinger equation. Through the interim paths between the end points, the paths have no substance and no interaction with external events. [4]

For the theory defined here, it is presumed that there is actually a probability of density of photons on those paths that interact with the probability density of the photons from other particles. The interaction of these probability densities is shown to explain phenomena such as electricity, gravitation, and the behavior of light without the existence of energy carrying fields.

A particle-based probability density is being proposed as the basis of an extended reformulation of the Feynman version of QED. [1] The following is an outline of the methodology.

Photons and Rest Mass in Δc Mechanics

The fundamental assumption of Δc Mechanics is that the universe is a construction of photons. The photon being a Planck particle having a velocity of c; if they are not spinning, they have a radius of about 10^{-33} cm and, if spinning, has an energy of $\hbar\omega$ and a Compton radius of $\lambda = c_0 / \omega$. The presumption is that mass is the local confinement of spinning polarized photons. Massive particles consist mostly of two photon self-bound pairs. The self-bound pairs are the primary leptons and quarks. Other massive particles, such as protons and mesons, are bound collections of those two-photon particles.

Concurring Phenomena

For any two photons, there is a single velocity frame in which the photons are equal and opposite, and, in that frame, the sum of the linear momentum is zero. The momentum and energy of two photons have a minimum energy velocity rest frame, and the energy and the momentum transforms the same as rest mass.

$$\frac{d}{dt}\left[(\omega_1 + \omega_2)\mathbf{r}_{CM} = \omega_1\mathbf{r}_1 + \omega_2\mathbf{r}_2\right] = 0 \tag{40}$$

Trapped photons, whether in a box or as self-bound pairs, exhibit the same momentum as a massive particle with equivalent energy.

A volume of random photons within a confined space could be self-bound in pairs, or free within a black body (Fig. 1).

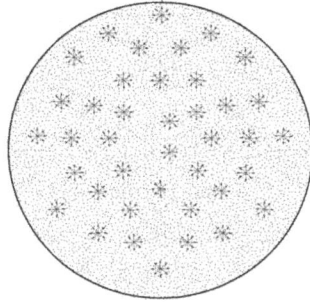

Random Collection of Photons

Fig. 1

The kinetic energy of each photon is $\varepsilon = \hbar\omega_n$, and it has inertial mass defined by its momentum; thus $p = mc$, and the value of its inertial mass is:

$$m = \sum_n \hbar\omega_n / c^2 \qquad (41)$$

When bound together, the sum of these photons constitutes the inertial mass of a rest particle, such as the electron. The momentum is the sum of the inertial mass of the photons times the velocity of the center of mass.

The total energy momentum and energy of a pair of photons from center of mass coordinate is:

$$p = M_n v_{CM} = (m_1 + m_2)v = \frac{\hbar\omega_1}{c_0^2}c_1 + \frac{\hbar\omega_2}{c_0^2}c_2 \qquad (42)$$

The total energy and momentum of a collection of random photons equate to those of a collection of massive particles with the same energy. Thus, it can be asserted that rest mass represents the local confinement of the kinetic mass of a collection of photons, supporting the notion that this is a physically plausible concept.

The difference in Feynman's view of QED and the presentation here presumes that the Feynman action paths are not empty but possess a probability of the photon actually being on the path. This probability introduces a particle-particle interaction by assuming that the fundamental particle is the photon, and in this framework, interactions between photons are mediated through the probability density of their positions on the action paths. The encounters of probability density from two sources reduce the velocity of light or increase the index of refraction for the encountering photons. [3]

Consider the action paths of a photon moving from one point to another along a line connecting the points, Fig. 2:

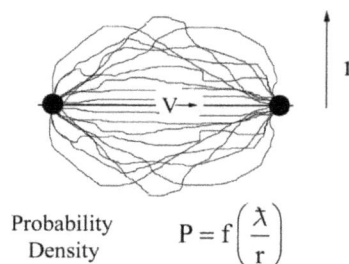

Probability Density $\quad P = f\left(\dfrac{\lambda}{r}\right)$

Fig. 2

The Feynman action sum starts with the initial Schrödinger state and reproduces it at the end, but between the points, the state or the particle doesn't exist. The view here is that photons traveling from point A to point B follow all possible paths. The probability density of their flow, P is highest near the primary path and inversely proportional to the distance from it, but the total probability of the existence of the photon is present at all positions on the line of travel.

$$P = f\left(\frac{\lambda}{r}\right) \tag{43}$$

As a photon traverses all possible paths between two points, not just the primary path, the probability existing on these paths interacts with the probability density of other particles. This interaction results in a decrease in the velocity of light for the action paths of both particles, mutually deflecting the action paths of the other particle in its direction (Fig. 3).

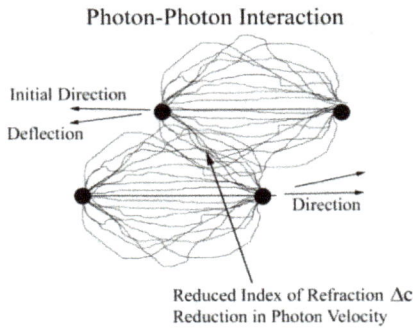

Photon-Photon Interaction

Initial Direction

Deflection

Direction

Reduced Index of Refraction Δc
Reduction in Photon Velocity

Fig. 3

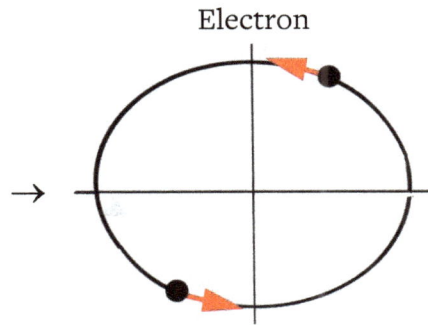

Electron

Fig. 4

Electron $\hbar\omega_p = 2 \times 0.25549947$ Mev

At a certain energy level of the photons $\varepsilon = \hbar\omega$, this deflection is sufficient to induce the polarized photons into a circular orbit, which manifests the binding as an electron (Fig. 4).

Photons moving in a self-bound orbit within the Compton radius generates a probability flow density outside this orbit, which can interact with similar particles. This interaction can be recognized as an electric field (Fig. 5). [2]

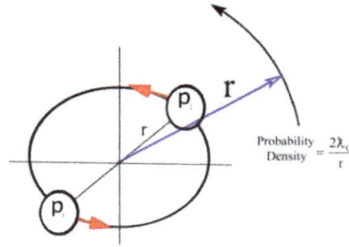

Fig. 5

As electrons and positrons engage, the same phenomena bind the particles together as shown in Fig. 6:

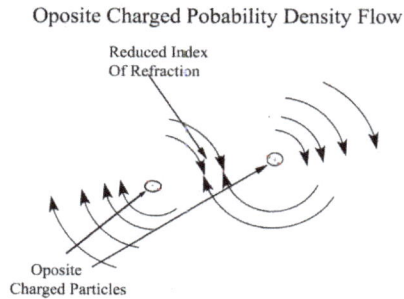

Fig. 6

From this model, one can derive gravitational and electric interactions, as well as the fine structure constant. The change in the velocity of light as a function of distance from the particle is given by:

$$\frac{\Delta c}{c_0} = \frac{1}{2}\left[\left(\frac{\sqrt{2}\lambdabar_{PL}\nu_e}{\lambdabar_e}\right)\left(\frac{\lambdabar_e}{r}\right)\right]^2 = \frac{1}{2}\left(\frac{\alpha\lambdabar_e}{r}\right)^2 \rightarrow \frac{1}{2}\left(\frac{\alpha}{n}\right)^2 \qquad (44)$$

122

The probability flow of two photon particles is planar, and the interaction between similar particles is in counter-rotating planes. The interaction between the photons repeats on each revolution, thus, multiplying the probability of interaction by the Compton frequency v_e^2 this makes about 10^{36} times the energy change induced by gravitation. [3]

Gravitation

The internal photons inside all mass particles generate probability flow, but because of the lack of a specific orientation, the flow direction exterior to the Compton radius is random and thus, the total exterior flow probability density direction is in random directions and spread over 4π radians of space. This generates a random flow of probability density that changes the velocity of light for interloping photons. Since it is ubiquitous in direction and not repeating at the Compton frequency at a point, it is about 10^{36} times smaller than the electric interactions.

The change in the velocity of light for a passing photon in proximity to mass particle is the probability of hitting a Planck particle inside the Compton radius of that particle times the probability of that particle being at the distance of the passing photon [3] is:

$$\frac{\Delta\varepsilon}{\varepsilon_0} = \frac{\Delta c}{c_0} = \left(\frac{\lambda_{PL}^2}{\lambda_p^2} \times \frac{2\lambda_p}{r} \right) = \frac{G\hbar}{c^3}\frac{mc}{\hbar\, r} \quad \rightarrow \quad \frac{2GM}{c_0^2 r} \tag{45}$$

This is the change in c as the result of the test particle moving from a great distance to the proximity of the mass M. The mass M is the sum of a group of particles at the center at $r = 0$. $\Delta\varepsilon$ is the change in the energy of a particle with energy ε_0 as a result of being at a distance r from the p particle with Compton radius λ_p. This is the expression that shows gravitation to be the combined probability effect of a group of single quantum particles, rather than a continuous field. Note that the Feynman photon, which is defined as a Planck particle without a Compton radius, has a cross section of about 10^{-66} cm, or about 39 orders of magnitude smaller than a neutrino.

Eq. (45) is the change in c, Δc, for a test mass particle m with energy ε_0.

$$\Delta\varepsilon = mc_0^2 - m'c_0^2 = \frac{2GM}{c_0^2 r}mc_0^2 \qquad (46)$$

This is the change in c, or the change in energy of the particle as it moves from a distance to a position r near the mass M. The escape energy for such a particle is:

$$\varepsilon = \frac{MG}{r} = \frac{1}{2}v_{esc}^2 \qquad (47)$$

Thus, the total change for the mass particle is a change in the rest mass at the position and the velocity:

$$m'c_0^2 = m_0 c_0^2\left(1 - \frac{GM}{c_0^2 r} - \frac{1}{2}v_{esc}^2\right) \qquad (48)$$

The escape energy can be dissipated, thus, leaving the rest mass at the "r" position as:

$$m_0 c_0^2 = m_o' c_0^2\left(1 + \frac{GM}{c_0^2 r}\right) \qquad (49)$$

Note that the dissipation of the kinetic energy leaves a rest mass less than the free mass, m_0, and unable to escape.

Combined Relativistic Velocity and Gravitational Attraction

A test particle moving in a proximity to a mass particle has a change in its rest mass, due to the decrease in the velocity of light, and a change in its observed mass due to the relative velocity of light.

Eq. (50) gives the relation between the rest mass and the relativistic velocity mass, and Eq. (49) and (51) gives the relativistic gravitational mass independent of its velocity.

$$m_0 c_0^2 = m_v' c_0^2 \left(1 - \frac{1}{2} \frac{v^2}{c_0^2} \right) \tag{50}$$

$$m_0 c_0^2 = m_o' c_0^2 \left(1 + \frac{GM}{c_0^2 r} \right) \tag{51}$$

The relativistic velocity rest mass of Eq. (50) is the relativistic proximity mass of gravitation Eq. (51) due to the presence of a gravitating mass, thus the two expressions can be combined as:

$$m_0 c_0^2 = m' c_0^2 \left(1 + \frac{GM}{r} \right) \left(1 - \frac{1}{2} \frac{v^2}{c_0^2} \right) \tag{52}$$

This forms the relativistic Lagrangian, which is the same as the same as Eq. (48) when the velocity is equal to the escape velocity.

$$\varepsilon_0 = m' c_0^2 \left(1 + \frac{GM}{c_0^2 r} - \frac{1}{2} \frac{v^2}{c_0^2} \right) \tag{53}$$

125

Conclusion

The concept of a photon-based QED, such as has been described here as Δc Mechanics, can provide the mechanical basis of Electric, Gravitational, and Velocity particle-particle mass interactions without reference to an energy density field.

References

1. DT Froedge, *The Physics of Delta-c Mechanics*, ISBN-13: 979-8218347178, Feb. 14, 2024, (Papers referenced can be found as chapters in this publication, and available online in DOI numbers), The primary reference for the developed items are in the cited papers.

2. DT Froedge, "The Electron as a Composition of Two Vacuum Polarization Confined Photons," April 2021, DOI: 10.13140/RG.2.2.18971.18722, https://www.researchgate.net/publication/350740864.

3. P. Halpern, *The Quantum Labyrinth: How Richard Feynman and John Wheeler Revolutionized Time and Reality*, Basic Books, New York, 2017.

4. R. Feynman – Nobel Lecture: The Development of the Space-Time View of Quantum Electrodynamics, 1965 https://www.nobelprize.org/prizes/physics/1965/feynman/lecture/.

Acceleration Gravitation and Origin of Centrifugal Force in Delta-c Mechanics and the Basis of E=mc2

D.T. Froedge

V070324

Introduction

Δ c **Mechanics** is a particle-based theory without the concept of fields, based on the effects of the density of Feynman photons, which are the probability of the existence of particles exterior to the Compton radius on action paths generated by photons in the interior of particles.

Δ c **Mechanics** shows that inertial mass particles are formed by the self-binding of photons in circular motion. The electron, muon, and tauon are creations of mass particles with two bound photons. The more complex particles, mesons, and protons are compositions of two and three of such particles. The effects of charge and gravitation are shown to be the action path interaction mechanisms of the photon probability density in changing the index of refraction, and not the effect of a field. [5]

Most of the previous work has focused on atomic and nuclear interactions, but the concept is general applying to the kinetics of mass velocity, acceleration, and gravitation, as well as the origin of the Lorentz transforms. This paper is to focus on the macroscopic aspects of the theory, particularly acceleration, relative velocity, and Lorentz transforms.

The differences between what is defined as Feynman vs Ordinary photons are the probability of its location.

All photons and Feynman photons exist as Planck (10E-33 cm) particles that move at the velocity of light. Ordinary photons are rotating particles, with their location probability density mostly inside the Compton radius. The same is true for self-bound photons. The energy of the photon is $\hbar\omega$ and the angular momentum is $\hbar$. The photons primary flow density is the movement of the particle inside the Compton radius, and the integral of the angular momentum of its probable flow density is the spin, $\hbar$.

A Feynman photon is the same particle but is distinguished by being on the action path outside the Compton radius and is not rotating. The integral of the angular momentum about the Compton radius of its flow density is the anomalous spin. The photon rotates about the central core but can be very far from the particle and thus does not transfer the energy or momentum of the particle. Its cross section can scatter and change the direction of other Feynman photons. The effect is by way of a change in the index of refraction, directing the motion of other photons and thus the motion of other interacting particles. Electric effects are created by differential rotational motion of internal particles, and there is no electric field or electric charge. [5]

Feynman's theory of quantum electrodynamics is based on the probability of action paths existing throughout space, but Feynman and Wheeler missed the point that the probable presence of particles on these action paths could have an effect on other particles. Δc Mechanics presumes that the presence of this interacting probability density is responsible for the effects attributed to fields, and no continuous energy fields exist.

For the proton, the change in the velocity of light in its proximity due to the photon probability flows of photons inside the proton.

$$\frac{\Delta c}{c_0} = \left(\frac{\lambda_{PL}}{\lambda_P}\right)^2 \frac{2\lambda_P}{r} \tag{1}$$

(See Appendix II and [8].)

The Compton radius of the Planck particle is λ_{PL} and for the proton is λ_P, the first term is the ratio of the area of the Planck particle to the proton, and the second term is the probable density of a Feynman photon being at a distance r from the particle. $\Delta c / c_0$ is the ratio of the change in the velocity of light to c_0 at that distance. Since protons are the primary constituents mass, this relation can be integrated over all the protons in the universe to get the background vacuum density of Feynman photons. This is about ($1.6E{+}38$ cm^{-2} sec^{-1}), and sets the speed of light in the universe.

Order of Presentation

Lorentz Transform
The Limit on the Speed of Light
Photon Mean Free Path

Lorentz Equivalence of △c Mechanics
Lorentz Postulates
Equivalence of the Lorentz Transforms
Moving Point (Part 1)
Constancy of c in Relative Moving Points (Part 2)
Lorentz Summary

Acceleration
The Origin of Centrifugal Force
The Energy of Circular Motion
Gravitation
   The Combined Relativistic, Velocity, and Gravitation Mass
Observations
Conclusion

References:

Appendix I   *The flow density of Feynman photons*
Appendix II   *The Electron as Two Orbiting Photons*
Appendix III   *The nature of the force between these particles*
Appendix IV   *Issues regarding △c Mechanics*

The issues dealt with in this paper are:
1.   △c Mechanics illustrates the origin of the Einstein relativistic mass relation and the cause of the constancy velocity of light in any rest or moving frame. It inherently gives the results of the Lorentz transform without the concept of frame change.
2.   The origin of centripetal force and the relativistic mass of rotating particles.
3.   The origin of gravitational attraction as a gradient in the velocity of light, or the vacuum photon flux. Energy carrying fields are not necessary for mass interaction.

Lorentz Transform

The Limit on the Speed of Light

The speed of light is limited by the intervening Feynman photons in its action path. The assumption of Δc Mechanics is that the speed of light is limited by the oncoming Feynman photon flux in the vacuum density; otherwise, the velocity would be infinite. The vacuum density is generated by the action paths of the internal photons that make up the particle mass. The product of the volume density, n_d, times c_0 is the vacuum flow density in space in cm^2 sec and is constant:

$$\frac{c_0 n_d}{cm^2 \, sec} = 1.693E + 38 \qquad (2)$$

The speed of light is the distance a group of photons can travel in one second and arrive in sufficient numbers to trigger an event. It is not measured by the emission and detection of the arrival of a single photon.

The intervening Feynman photons between the emission and the arrival point determine the probability of a photon scattering out of the initial path, thus setting the mean free path and the number of photons available to trigger an event at a given distance. An increase in the intervening photon density increases the probability of the photon scattering out of the path, thus decreasing the detectable distance, the mean free path, and the velocity of light.

For convenience in this paper, one-second time steps, thus the above is the number of photons encountered in a second of travel. The product is a constant and, thus, higher density represents a slower velocity of light. Flow density n_d (cm-2 sec-1) is the number of photons in space with density of n_d cm^3 that would pass through an area of one cm in one second. (See Appendix I.)

The velocity is seen to be inversely proportional to the oncoming density, and the product of the velocity of light times the oncoming flow density is a constant per time step (sec).

The product of the density times an altered velocity, c', can be equally viewed as a change in the density because $n_d c' = n_d' c_0$.

The mean free path of a photon, S, encountering a density of $n_d \, / \, cm^2 \, sec$ Feynman photons and having a intersecting radius between two particles of 1.5 times the Planck radius is:

$$\frac{n_d \, c}{cm^2 \, sec} \frac{(1.5\lambda_{PL})^2 \, cm^2}{} \frac{S \, sec}{c} = \frac{1}{2} \quad \rightarrow \quad S = \frac{1}{2} \frac{1}{n_d (1.5\lambda_{PL})^2} = 5.02E+26 \, cm$$

(3)

Lorentz Equivalence
of Δc Mechanics

Lorentz Postulates

The Lorentz transform is based on two assumptions put forth by Einstein. Those assumptions are true, but neither define the underlying mechanism for these assumptions or provide a theoretical basis.

The first assumption defines the inertial mass relation between stationary and moving frames:

$$m' = \frac{m_0}{\sqrt{1-v^2/c^2}} \quad \rightarrow \quad m'^2 c_0^4 - m'^2 v^2 c_0^2 = m_0^2 c_0^4 \quad \rightarrow \quad \varepsilon^2 c_0^4 - p^2 c_0^2 = m_0^2 c_0^4$$

(4)

This can be written as:

$$\left[m'(c_0 - v)c_0 \right] \cdot \left[m'(c_0 + v)c_0 \right] = m_0^2 c_0^4 \qquad (5)$$

Or more compactly with Dirac matrix in Geometrical Algebra as:

$$p = m'(\gamma^k c_0 + \gamma^0 v) \quad \rightarrow \quad \frac{\varepsilon^2}{c_0^2} = p^2 = m_0^2 c_0^4 \qquad (6)$$

Although this assumption is provably correct, it does not contain the physical underlying mechanics.

The second Lorentz postulate asserts constancy in the speed of light in any frame, moving or stationary:

$$c_1 = c_2 \tag{7}$$

The Lorentz transforms give the mathematical structure to this by transforming four Minkowski coordinates of an event into one inertial frame in terms of the coordinates of the same event in another inertial frame.

The coordinates (x, y, z, t) of an event in frame f_1 in terms of the coordinates (x', y', z', t') of the same event in frame f_2, are:

$$t = \frac{t' + vx'/c^2}{\sqrt{1 - v^2/c^2}}. \qquad x = \frac{x' + vt'}{\sqrt{1 - v^2/c^2}} \tag{8}$$

The Lorentz transform presumes equivalents of inertial frames and the constancy of the velocity of light to arrive at the transformation of the inertial properties for moving particles. This is also proven to be true experimentally but also does not explain the underlying cause.

Equivalence of the Lorentz Transforms

Moving Point (Part 1)

In Δc Mechanics, the velocity of light is set by the probable number density of Feynman photons confronting the photon as it passes, and the velocity of light is inversely proportional to the density of oncoming vacuum Feynman photons, $n_d c_0 = n'c'$, per cm² second. The concept of a frame change is not invoked, just the change in the density of the vacuum density as a result of the change in velocity.

Δc Mechanics starts with the presumption that mass is the local confinement of photons, primarily consisting of self-bound pairs.

The total mass of a body is the sum of the energy of all the photons scaled by the square of the velocity of light.

$$m = \hbar \left(\omega_1 + \omega_2 + \cdots \right) / c_0^2 \qquad \boldsymbol{p} = m\boldsymbol{c} \qquad |\boldsymbol{p}| = \hbar \left(\omega_1 + \omega_2 + \cdots \right) / c_0$$

(1.9)

For a stationary mass, m, the sum of the random momentum of all the internal photons of a mass in a given direction $\widehat{\boldsymbol{u}}$ is equal to the sum of the momentum in the opposite direction, this is the necessary requirement for the mass to be stationary:

$$\boldsymbol{p}_1 = m \sum \widehat{\boldsymbol{u}} \cdot \boldsymbol{c}_1 \quad = \quad \boldsymbol{p}_2 = m \sum -\widehat{\boldsymbol{u}} \cdot \boldsymbol{c}_2 \qquad (1.10)$$

This a constant in time and thus the product of these two sums is a time invariant

The magnitude of the product of these two momentum vectors of the internal photons is invariant and defines the scaler invariant energy of the mass.

$$\boldsymbol{p}_1 \cdot \boldsymbol{p}_2 = m\boldsymbol{c}_1 \cdot m\boldsymbol{c}_2 \quad = \quad \frac{\varepsilon^2}{c^2} \qquad (1.11)$$

This is the foundation of the Einstein relation $\varepsilon = mc^2$

Product Viewed by External Moving Observer

An observer moving with respect to the mass will see a difference in the magnitude of the two momentum vectors in the forward and reverse direction along the velocity.

Change in Arriving Density

For an observer in a stationary point, the number of particles moving through one cm² in one second is $n_d c_0$. For a moving point, the stationary observer sees a plus or minus increase in the density of photons arriving at the particle due to the velocity of the particle. (With no change of frame.)

This is illustrated in Fig. 1, as the particle moves with the photons from the left, there is a decrease in the photon density from the left arriving at the particle.

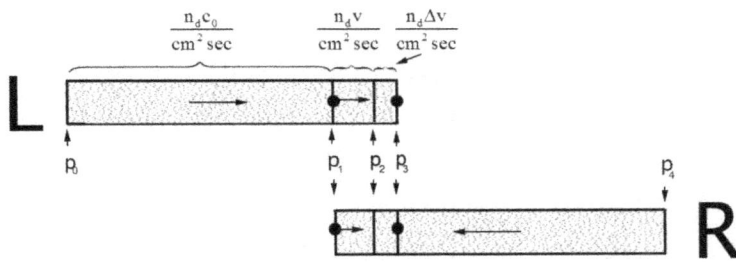

Feynman photon flux arriving at a moving mass viewed from a stationary observer

Fig. 1:

The distance from p_1 to p_2 is the distance c_0 travels in one second, thus the density of Feynman photons passing p_1 from the left is $n_d c_0$, and from the right is the same.

$$n_d c' = n_d c_0 - n_d v \tag{12}$$

If the particle is moving at v to the right, then when the particle arrives at p_2, the density passing the particle during this time from the left is reduced by $n_d c_0 - n_d v$, and during this same time the number of photons passing from the right is increased by $n_d c_0 + n_d v$. The velocities coming from the left and right can be set as c'_1 and c'_2, thus:

$$c'_1 = \left(c_0 + v\right) \;\; \rightarrow \;\; c'_2 = \left(c_0 - v\right) \tag{13}$$

From Eq. (13), the velocity of light observed, determined in the stationary frame, gives different values for the momentum vector, thus at the moving point:

$$p_1 = m(c_0 - v) \qquad p_2 = m(c_0 + v) \qquad (14)$$

Eq. (1.11) is then:

$$m^2\left((c_0 - v)\cdot(c_0 + v)\right) \;=\; m_0^2\,(c_0 \cdot c_0) \qquad (15)$$

This relation is the postulate of Einstein in Eq. (5); therefore, the theory being proposed provides the physical cause for the Einstein postulate of the Lorentz transform.

Observation from the stationary position creates an asymmetry in vacuum density, n_d, for the moving particle that changes the momentum vectors and value of the energy in the stationary frame. The basis of Δc Mechanics gives the same inertial relation as in Eq. (4) and Eq. (6), and yield the same relativistic mass in the observers position.

Writing this out shows the familiar relativistic velocity relation for a moving mass:

$$\varepsilon = m_0 c_0^2 = c_0\sqrt{m'^2\,c_0^2\left(1 - \frac{v^2}{c_0^2}\right)} = m'c_0^2\left(1 - \frac{1}{2}\frac{v^2}{c_0^2}\right) \qquad (16)$$

Changing the order shows the total energy of a moving mass to be the sum of the rest and the kinetic energy, where the kinetic mass is the relativistic mass:

$$\varepsilon_T = m'c_0^2 = m_0 c_0^2 + m'\frac{1}{2}v^2 \qquad (17)$$

Constancy of c in Relative Moving Points

(Part 2)

In Δc Mechanics, the velocity of Feynman photons at any point in space from all directions is equal and symmetric because the flow density is presumed to be in thermodynamic equilibrium.

Figure 2 is a sketch of an observer or particle at rest, and an observer moving with a velocity v.

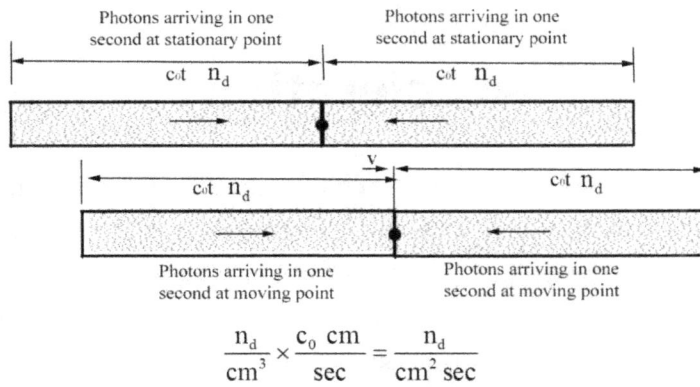

$$\frac{n_d}{cm^3} \times \frac{c_0 \; cm}{sec} = \frac{n_d}{cm^2 \; sec}$$

Fig. 2: Speed of light in two moving frames

The velocity of light at a point is set by the opposing density of Feynman photons, and since the background density moves in all direction at c_0 , the number of photons per cm^3 is the same no matter what direction or velocity. All the photons coming from a distance of 29979245800. cm still arrive in one second, and photons coming from the opposite directions at a distance of 29979245800. cm come thru in one second. The space density of Feynman photon flow is observed to be homogeneous.

Thus, from observation, the theory in defining the velocity of light assures that the velocity of light in all moving frames is the same. This gives cause to the second Lorentz postulate.

The section (part 1) observes the moving plane from a stationary frame yield the mechanism of the Einstein presumption, causing the relativistic mass Eq. (4),

The section (part 2) shows the basis for the equivalence of the velocity $c_1 = c_2$ for observers in separate moving frames.

Thus, Δc Mechanics provides an alternate to the Lorentz transforms and shows the physical cause underlying both base postulates.

Acceleration

The previous section discussed the change in velocity of light in a moving frame, but there is also a change due to a change in the velocity, or the acceleration shown in Fig. 1. The block from p_2 to p_3 is the change in velocity from the time photons at p_0 move to p_3, which is the integral of the acceleration for one time step or one second. This is the integral of the increase in velocity from the start at p_1 to the arrival at p_3. Eq. (12) thus becomes:

$$n_d c' = n_d c_0 + n_d v + \int_{v=o}^{v=1sec} n_d a\, dt \qquad (18)$$

This change in the density of approaching photons is a change in the velocity of light experienced by the particle and thus, including the acceleration in Eq. (15) becomes:

$$m_0^2\, \mathbf{c}_0 \cdot \mathbf{c}_0 \;=\; m^2 \left[\mathbf{c}_0 - \left(\mathbf{v} + \int_0^{1sec} \mathbf{a} dt \right) \right]\left[\mathbf{c}_0 + \left(\mathbf{v} + \int_0^{1sec} \mathbf{a} dt \right) \right] \qquad (19)$$

The acceleration is a vector and, if in the same direction of the velocity, is just the last second of acceleration, but it is not necessarily in the same direction as the velocity.

The Origin of Centrifugal Force

The origin of centrifugal force has been a historic mystery since Newton developed the concept of force. Mach's view, that the origin of centrifugal force is a result of the relative motion of the distant stars was initially the presumption of Einstein in developing GR, and it still may be true, but locally this not the mechanism.

A deviation from a straight line does not necessarily change the magnitude of the velocity, but it does require energy to be applied, $\Delta\varepsilon = f \cdot dr$. When a central force is applied to a rotating object, as the force is increased, the force leads the velocity and changes the velocity of the mass. But when a purely lateral force, to the velocity, is applied, the velocity remains constant, but the force is still providing an increase in energy to the system, $\Delta\varepsilon = f \cdot dr$.

The Energy of Circular Motion

The answer to the mystery of centrifugal force is that the relativistic energy of a mass moving in a circle is greater than that of the same mass moving in a straight line. The lateral acceleration of a particle requires energy that doesn't increase the velocity but increases a hidden relativistic mass. When the acceleration stops, the particle returns to a straight line and the centrifugal energy must be radiated away as gravitational radiation.

Thought Experiment

Consider a thought experiment in which a mass is rotating in a circle with velocity $v_m = v_i$ (Fig. 3):

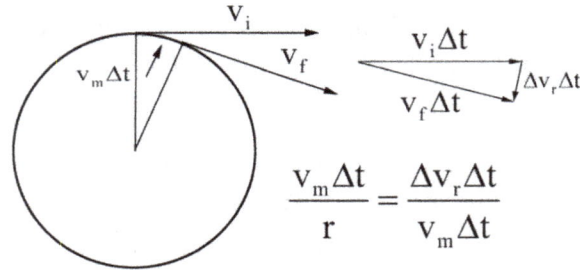

Fig. 3: Rotating particle

After a short interval of time the angle between the radii positions and the velocity vector form similar isosceles triangles, and here is equality between the ratio of the base and one leg.

$$\frac{v_m \Delta t}{r} = \frac{\Delta v_r \Delta t}{v_i \Delta t} \qquad v_i = v_m \qquad (20)$$

The centripetal acceleration is $a = v_m^2 / r$, and with some manipulations Eq. (20), becomes:

$$v_m^2 \Delta t^2 = r \Delta v_r = r(a_r \Delta t) = r \frac{v_r^2}{r} \Delta t = v_r^2 \Delta t$$
$$v_m^2 \Delta t^2 = v_r^2 \Delta t \qquad (21)$$

From the section on acceleration, the time interval for the density flow is one second, thus integrating over one second gives the radical change in the density of photons in that second, which is equal to the radial velocity change due to the acceleration.

$$n_d \int_{t=0}^{t=1} v_m^2 \Delta t^2 = n_d \int_{t=0}^{t=1} v_r^2 \Delta t \rightarrow n_d v_m = n_d v_r \qquad (22)$$

The number density of photons approaching the mass along the radius is reduced the same as from the lateral direction. Eq. (5) becomes:

$$m_0 c_0^2 = mc_0^2 \left[1 - \frac{1}{2c^2} (\mathbf{v_m} + \mathbf{v_r}) \cdot (\mathbf{v_m} - \mathbf{v_r}) \right] \qquad (23)$$

The vector velocities of $\mathbf{v_m}$ and $\mathbf{v_r}$ are perpendicular, and thus their dot product vanishes, leaving:

$$mc_0^2 = m'c_0^2 \left[1 - \left(\frac{1}{2} \frac{v_m^2}{c_0^2} + \frac{1}{2} \frac{v_r^2}{c_0^2} \right) \right] \qquad (24)$$

And the total relativistic energy is:

$$\varepsilon = m'c_0^2 = \left(mc_0^2 + \frac{1}{2} m'v_m^2 \right) + m'\frac{1}{2} v_r^2 \qquad (25)$$

The relativistic energy is the rest energy plus twice the kinetic energy. The term in parenthesis is the relativistic energy of the moving particle, Eq. (17). The second term is the additional relativistic mass created by circular acceleration. When the acceleration ceases (i.e., cutting the string), the energy vanishes and thus must be radiated by the change in acceleration or jerk, presumed to be by gravitational radiation.

Summary

This change in the relativistic mass in Eq. (25), due to circular motion, shows up in Δc Mechanics, but 0n10ot in the Lorentz transforms, because of the Einstein presumption of Eq. (4) does not include the change in the relativistic mass due to an acceleration and is not present in a four-space coordinate transform.

It has been shown that centripetal force is more a local phenomenon related to travel in straight line than the motion of the distance stars, and it is found that there is energy hidden in the relativistic mass.

Since this circular centrifugal energy does not contribute to a change in the magnitude of the velocity, it is a hidden contribution to the relativistic mass, and whether it can be detected by experiment is an interesting question.

Gravitation

Gravitation is a particle generated phenomenon caused by the change in c induced by the Feynman photons on action paths exterior to the Compton radius. The self-bound photons revolving in the proton generate most of the probability density of Feynman photons, creating the gradient in c that is the source of gravitation. Gravitation is not generated by a mass density of space.

Except for the low-mass electron there is no particular orientation of the rotating photons in the proton, thus the probable exterior action flow density of the Feynman photons have no particular direction. The flow density exterior to the proton Compton radius exists as a random directed density of probability flow.

The acceleration by mechanical or electric interaction takes place in an environment of a constant vacuum density of Feynman photons, but the acceleration due to gravity is from a gradient in the vacuum density locally producing a gradient in the velocity of light.

Specifics

The change in the velocity of light for a proton was arrived at in [8] based on the intersecting probability density of Feynman photons from the proton intersecting with the vacuum density of space. As expressed here, the change in the velocity of light in the proximity of a proton is:

$$\frac{\Delta c}{c_0} = \left(\frac{\lambda_{PL}}{\lambda_P}\right)^2 \frac{2\lambda_P}{r} \tag{26}$$

(See Appendix II.)

The radius of the Planck particle is λ_{PL}, and the Compton radius of the proton is λ_P. The term in parenthesis is the probability of hitting a Planck particle within the Compton radius of a proton, and the last term is the probability of a Planck particle from the proton being at a distance r from its center, constituting the Feynman photon density.

This vacuum density is the basis of the electron creation radius $\mathfrak{R}_0$, and the mass of the electron, thus providing the connection between gravitation and quantum effects. [5, 8]

Combining Relativistic Mass Effects

Gravitation changes the Feynman photon density n_d and the velocity of light. The operable relations in Δc Mechanics, Eq. (27), below are: (a) the change in c due to the probability density in the presence of a single proton, Eq. (1); (b) the Blandford, Thorne velocity of light vs gravitation arrived at Solar delays and solar bending, in a Minkowski space, [15, 17], and; (c) the estimated number of proton equivalent mass particles in the universe. [16]

$$(a)\frac{\Delta c}{c_0} = \left(\frac{\lambda_{PL}}{\lambda_P}\right)^2 \frac{2\lambda_P}{r} \qquad (b)\ \frac{c}{c_0} = \left(1 - \frac{2\mu}{r}\right) \qquad (c)\ n_P = \frac{R}{2\mu_P}$$

(27)

143

In Eq. (27), μ is the gravitational radius, R is the radius of the universe, and n_p is the number of proton equivalent mass in the universe.

The Combined Relativistic, Velocity and Gravitation, Mass

A test particle placed in a proximity to a mass particle has a change in its relativistic mass, due to the decrease in the velocity of light, Eq. (27) (b), and the moving particle has a change in its relativistic mass due to the flow of the vacuum photon density, Eq. (17):

$$m_0 c_0^2 = m_v' \left(c_0^2 - \frac{1}{2} v^2 \right) \qquad (28)$$

$$m_0 c_0^2 = m_g' c_0^2 \left(1 - \frac{2\mu_P}{r} \right) \qquad (29)$$

This is the same test particle, thus the mass of expression, (51), m_g' is the zero velocity mass of expression (50), $m_g' \rightarrow (m_0)_v$, and the expressions can be combined to be:

$$m_0 c_0^2 = m' \frac{\left(c_0^2 - \frac{1}{2} v^2 \right)}{\left(1 - \frac{2Gm_P}{c_0^2 r} \right)} = m' c_0^2 \left(1 - \frac{1}{2} \frac{v^2}{c_0^2} + \frac{2\mu_P}{r} \right) \qquad (30)$$

This is a relativistic Lagrangian for a moving mass particle in a in a gravitational field:

$$m_0 c_0^2 = m' c_0^2 \left(1 - \frac{1}{2} v^2 + \frac{2Gm_P}{r} \right) \qquad m_P = \sum (n \text{ proton masses}) \quad (31)$$

Note that the gravitational term is not a negative field potential.

Observations

The constancy of the velocity of light in moving frames has been shown.

The effect of gravitation is caused by a change in the speed of light induced by the individual particles in a mass; there is no charge, no electric field, and no potential energy generated by gravitation.

From Eq. (31), if a test particle arrives from a great distance entering the presence of a mass having a kinetic energy equal to the gravitational energy, $\frac{1}{2} v^2 = \frac{2Gm_P}{r}$, this shows the relativistic mass does not change, and the velocity is constantly equal to the escape velocity. Should a particle in a gravitational presence lose its kinetic energy, the relativistic rest mass is then less than m_0, and it cannot escape without the addition of an escape velocity. If two equal mass particles come together and radiate the kinetic energy away, they are trapped with energy less than their free mass values.

In Gravitational motion the energy components of velocity and mass are being exchanged by a gradient in the density of Feynman particles. There is no negative energy supplied by a ubiquitous negative Gravitational field of unknown origin or location. There is in fact, no field at all.

Conclusion

The purpose of this paper is to illustrate the macro mechanics and utility of Δc Mechanics in physical phenomena that needed some clarification. Mass is nothing more than localized or self-trapped photons, $m = \hbar(\omega_1 + \omega_2 +)/c_0^2$, and the gravitational and electric effects are the result of the probability density of the Feynman action paths outside the confines of the Compton radius.

The speed of light is limited and set by the probability density of oncoming Feynman photons of the vacuum density generated by the internal photon action paths of mass particles. Gravitation is the gradient in the Feynman vacuum probability density generated primarily by a collection of protons.

Δc Mechanics illustrates the physical mechanism of the Lorentz transformation, acceleration, the centrifugal forces, electric, and gravitational interaction. There are no un-locatable field energy contents or excluded energy infinities of the standard view of QED.

Δc Mechanics is the genesis of a new formulation for physics, and this paper has been to present the theory in a more familiar context, for those familiar with general physics, but it is unlikely to create much notice or be accepted in the foreseeable future.

The Feynman action path formulation of QED yields an understanding of the underlying mechanisms of physics. Adding the particle view of the action paths, rather than a field concept, removes infinities and allows a physical understanding of the process. The processes shown here include: the velocity of light, particle-particle interaction, gravitation, and the velocity transforms.

References

1. DT Froedge, *The Physics of Delta-c Mechanics*, ISBN-13: 979-8218347178, Feb. 14, 2024, (Papers 2–14 and 19 can be found as chapters in this publication).

2. DT Froedge, "The Dirac Equation and the Two Photon Model of the Electron revised," April 2021, DOI:10.13140/RG.2.2.19095.70564, https://www.researchgate.net/publication/350922403.

3. DT Froedge, "The Connection between Electric Charge, Gravitation, and the Feynman Sum over All Histories View of Quantum Electrodynamics," April 2020 Conference: APS, April 18–21, 2020, Washington, DC, https://absuploads.aps.org/presentation.cfm?pid=18355, https://www.researchgate.net/publication/341310206.

4. DT Froedge, "A Quantum Theory Conjecture on the Origin of Gravitational and Electric Particle Interaction," December 2019, DOI: 10.13140/RG.2.2.29097.54884, https://www.researchgate.net/publication/337826826.

5. DT Froedge, "The Electron as a Composition of Two Vacuum Polarization Confined Photons," April 2021, DOI: 10.13140/RG.2.2.18971.18722, https://www.researchgate.net/publication/350740864.

6. DT Froedge, "The Gravitational Constant to Eleven Significant Digits," March 2020, DOI: 10.13140/RG.2.2.32159.38564, https://www.researchgate.net/publication/339943651.

7. DT Froedge, "The Fine Structure Constant from the Feynman Path Integrals," March 2021, DOI:10.13140/RG.2.2.12979.55846, https://www.researchgate.net/publication/350188862.

8. DT Froedge, "Vacuum Polarization, Gravitation, Charge, and the Speed of Light," Sept. 2021, DOI:10.13140/RG.2.2.15619.22569, https://www.researchgate.net/publication/354474157.

9. DT Froedge, "The Calculated value of the Fine Structure Constant from Fundamental Constants," September 2021, DOI:10.13140/ RG.2.2.34349.41440.

10. P. Halpern, *The Quantum Labyrinth: How Richard Feynman and John Wheeler Revolutionized Time and Reality*, Basic Books, New York, 2017.

11. R. Feynman - Nobel Lecture: The Development of the Space-Time View of Quantum Electrodynamics, 1965 https://www.nobelprize.org/ prizes/physics/1965/feynman/lecture/.

12. M Fowler, "Modern Physics," *Notes on Special Relativity Physics 252*, University of Virginia, March 21, 2008,

http://galileo.phys.virginia.edu/classes/252/SpecRelNotes.pdf, http:// galileo.phys.virginia.edu/classes/252/home.html.

13. K. Thorne, *John Archibald Wheeler 1911–2008 A Biographical Memoir*, https://arxiv.org/ftp/arxiv/papers/1901/1901.06623.pdf.

14. (PDF) "Neutrino Binding Between Nuclear Particles in Delta-c Mechanics Addendum to: Nuclear Particle Structure in Delta-C Mechanics," https://www.researchgate.net/publication/380397398.

15. Roger Blandford and Kip S. Thorne, *Applications of Classical Physics*, Chapter 25, http://www.pmaweb.caltech.edu/Courses/ph136/yr2012/

16. D. Valev, "Estimations of total mass and energy of the universe," arXiv:1004.1035v1 [physics.gen-ph] April 7, 2010.

17. I. Shapiro; Gordon H. Pettengill; Michael E. Ash; Melvin L. Stone; et al. (1968), "Fourth Test of General Relativity: Preliminary Results," *Physical Review Letters*. 20 (22): 1265–1269. Bibcode:1968PhRvL..20.1265S. doi:10.1103/PhysRevLett.20.1265.

18. DT Froedge, "Electron Mass and State Energy Levels Resulting from Photon-Photon Interaction," Conference: APS, April 2022, New York, https://www.researchgate.net/publication/359912763.

19. DT Froedge, "The Feynman Photon-Photon Electric Attraction, Repulsion and the Chirality of the Electron." DT Froedge, *The Physics of Delta-c Mechanics*, ISBN-13: 979-8218347178, Feb. 14, 2024.

Appendix I

The flow density of Feynman photons

The Local volume density of Feynman photons at a point is constant, and the flow density of photons from any two opposite directions Integrated over all directions are Equal, thus the integral of flow density for any point in any direction is the same.

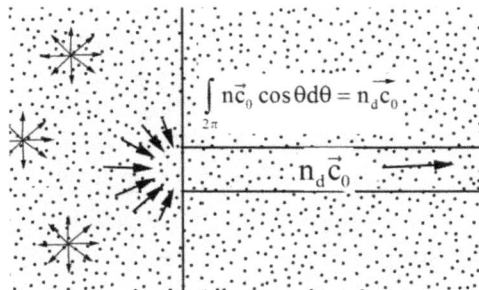

$$\int_{2\pi} n\vec{c}_0 \cos\theta \, d\theta = n_d \vec{c}_0$$

$n_d \vec{c}_0$

Fig. 4

Figure 4 is a sketch of space with a volume density of Feynman photons integrated over the 2 pi solid angle.

The Flow density n_d (cm^{-2} sec^{-1}) is the integral of the number of photons in space with density of n_d cm^3 moving at c that would pass through an area of one cm in one second. (Appendix I)

A photon moving in space encounters an opposite flow of Feynman photons of about 1.6932057E+38 photons per second per square cm. With that opposing density, the photon can travel about 2.9979245800E+10 cm in one second. The integral of photons going in a particular direction over a given area is the same, and each photon experiences the same oncoming density. The number and flow density are homogeneous.

Appendix II

The Electron as Two Orbiting Photons

Excerpts with corrections from:

Electron Mass and State Energy Levels Resulting from Photon-Photon Interaction [18]

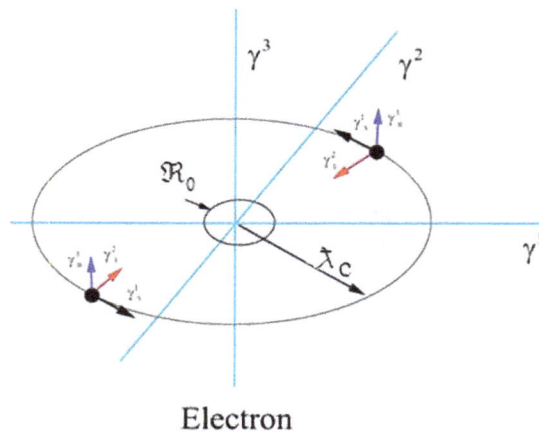

Fig.1: The vector orientation of the probability of location of two radially polarized photons rotating in the electron at the first most probable $L=\hbar$, orbit with r equal to the electron Compton radius. Red is the electric vector, Blue is the magnetic vector, and Black is the Poynting vector.

Electron

Mechanics of the Electron

The electron is a composition particle composed of two photons, each with half the energy of the electron, revolving around the center of mass at a probable distance of the Compton radius. A photon passes or engages the probability of the other photon twice per revolution, thus the frequency of photons passing the other photon is twice the rotation frequency. For a fixed point, there are two photons passing per revolution that provides the proper Compton frequency of the electron. The Compton frequency is twice the rotation frequency of the photons.

From earlier papers, [2–9], the change in the velocity of light of a passing photon as the result of the probability of another photon passing at a perpendicular distance r is:

$$\frac{\Delta c}{c_0} = \frac{\lambdabar_{PL}^2}{\lambdabar_{ph}^2}\left(\frac{\lambdabar_{ph}}{r}\right) = \left(\frac{\lambdabar_{PL}^2}{\lambdabar_{ph}^2}\right)\left(\frac{\hbar}{m_{ph}cr}\right) \qquad (32)$$

$$(\text{cross-section ratio}) \times \text{Feynman photon probability density}$$

The first term in brackets is change in the density of Feynman photons as the result of encountering the "ph" photon, and the second is the probability of the "ph" photon passing at a perpendicular distance r from the observed point. The momentum of the photon is $m_p c_0$, the Planck particle radius is $\lambdabar_{PL}$, and the Compton radius of the photon is $\lambdabar_{Ph}$.

If a photon is bound in orbit, at a frequency per revolution of v_1, an interloping photon encounters the universal background density of Feynman photons per second [2–4], plus the density of the rotating photon. This slows the velocity of light proportional to the frequency.

$$\frac{\Delta c}{c_0} = \frac{\lambdabar_{PL}^2}{\lambdabar_1^2}\left(\frac{\lambdabar_1}{r} v_{ph}\right) \qquad (33)$$

The other photon in the electron is not an interloping photon but participating. The velocity of the photons is proportional to the density of Feynman photons in the universe, plus the density of encountering the other photon. The change is proportional to the probability of encountering the other photon, thus the "coincidence" of the probable encounter of the photons in an orbit.

The product of the probability of the photon densities and ,thus, the "coincident" probability of collision is:

$$P_1 P_2 = \frac{\lambdabar_1}{r} v_{ph} \frac{\lambdabar_2}{r} v_{ph} = v_{ph}^2 \frac{\lambdabar_1}{r}\frac{\lambdabar_2}{r} \qquad (34)$$

The probability densities expressed here are for the Feynman photons of photons passing at a distance. The square of the frequencies, v_{ph}^2, is the coincident collision frequency of the photons

The frequency of rotation is v_e, thus the distance the photon travels at this distance in one second at the core of the electron, c_e, is $S = 2\pi(\lambdabar_{PL} 2v_{ph})v_e$. The ratio of this distance to the distance light moves in one second is:

$$\frac{c}{c_0} = \frac{(\lambdabar_{PL} v_e) \times 2\pi v_e}{c_0 t} = \frac{\Re \omega_e}{c_0 t} \qquad (35)$$

The radial value, $\Re$, is designated as the "Electron Creation Radius," since it represents the radius at which the potential energy of the two rotating photons is equal to the total energy. The maximum of the potential energy can only be half the total energy at the point at which it is equal to the kinetic energy, thus we will define:

$$\Re_0^2 = 2\lambdabar_{PL}^2 v_e^2 \qquad (36)$$

This is the core radius of the electron; it is an invariant quantity that defines the invariant mass. Two photons cannot actually be closer than this because the kinetic energy of the photons, p, would be less than the invariant energy.

$$\frac{\Delta c}{c_0} = \frac{\lambdabar_{PL}^2}{\lambdabar_1^2}\left(\frac{\lambdabar_1}{r} v_{ph}\right) \qquad (37)$$

The other photon in the electron is not an interloping photon, but participating. The velocity of the photons is proportional to the density of Feynman photons in the universe, plus the density of encountering the other photon. The change is proportional to the probability of encountering the other photon, thus the "coincidence" of the probable encounter of the photons in an orbit.

Appendix III

The nature of the force between these particles

Excerpts from:

The Feynman Photon-Photon Electric Attraction, Repulsion and the Chirality of the Electron [19]

Opposite Charges

For a simple atomic system, Positronium, there are two opposite, rotating particles. Each particle is surrounded by a circulating probability of the core photons being in a planar rotating path. As shown in Fig. 2, the photons from both particles rotate and engage the photons from the other particle.

Attraction or repulsion between the particles depends on direction of the interacting photons. This is the result of the Lorentz condition that photons in the same direction cannot interact, whereas opposite going photons have a probability of collision and reduce each other's relative speed of light, or index of refraction.

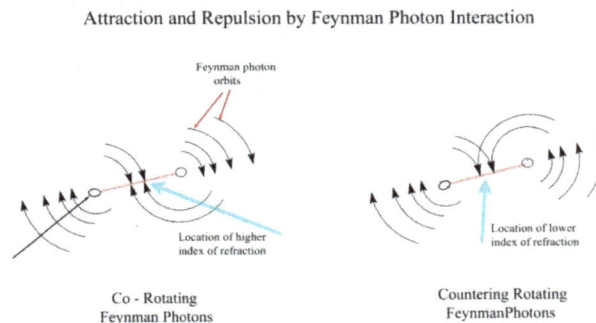

Attraction and Repulsion by Feynman Photon Interaction

Feynman photon orbits

Location of higher index of refraction

Location of lower index of refraction

Co - Rotating Feynman Photons

Countering Rotating FeynmanPhotons

Fig. 2: Interacting rotating particles

The collision probability in Eq. (34), and the magnitude of the change in c, Eq. (37), as the result of a charged particle at a distance r:

$$\frac{\Delta c}{c} = \frac{1}{2} \frac{1}{m_e c^2} \frac{\alpha c \hbar}{r} \qquad (38)$$

Δc is the change in the velocity of light or the index of refraction.

Appendix IV

Issues regarding Δc Mechanics

Planck vs Ordinary Photons

Ordinary photons are rotating Planck particles with their probability of location on a rotating line inside the Compton radius from the perimeter through the center and back. The energy of the photon is $\varepsilon = \hbar\omega = h\nu$, and the angular momentum is $\hbar$. The photons flow density is the movement of the particle inside the Compton radius, and the integral of the angular momentum of its flow density is the spin.

A Feynman photon is the probability of the particle's location on action paths outside the Compton radius, and the integral of the angular momentum of its flow density is the anomalous spin. The Feynman photon can be miles away from the particle and thus it cannot transfer energy or momentum, but its cross section and velocity can scatter and change the direction of other Feynman photons.

Feynman's theory of quantum electrodynamics is based on the probability of these paths being out there. Feynman did not presume that the probable presence of particles on these action paths had any effect on other particles, but Δc Mechanics presumes it is the interaction mechanism between mass particles rather than that of continuous fields. Changing the index of refraction experienced by the Feynman photons of another particle changes its direction and thus can attract or repel other particles.

Relativistic Time Dilation Illustrated in Delta-c Mechanics

D.T. Froedge

V120923

Abstract

This paper shows the Δc Mechanics model of the electron, by virtue of its structure, demonstrates time for mass in a moving frame exists at a slower rate than time for the electron in the stationary frame.

The electron, as shown in Δc Mechanics, is a pair of self-bound photons rotating around the center of momentum. [1, 2] All electrons have the same mass and Compton frequency, which is twice the Compton frequency of the individual rotating photons.

~

The electron has spin, so for simplicity the rotation axis can be aligned along its direction of travel. This configuration is useful in demonstrating the difference in the rate of the passage of time.

The time of rotation, or period, for an observer in the moving frame is:

$$\Delta t_M = \frac{2\pi r}{c_M} \tag{1}$$

The time of rotation in the stationary observer's frame in which the velocity of light, c_0, is:

$$\Delta t_M = \frac{2\pi r}{c_M} \qquad (2)$$

A sketch of the motion of photons in the electron illustrates the spiral structure of the trajectory in the stationary frame (Fig. 1).

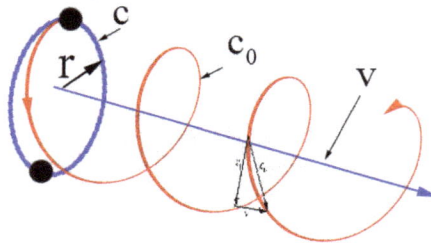

Fig. 1

For the photon to make a revolution around the center of mass for an observer in the rest frame, the photon travels at c_0, but the distance is further than for an observer of the photon in the moving frame. Pictographically, this is:

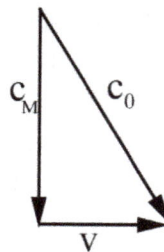

And from the Pythagorean Theorem:

$$c_M = c_0 \sqrt{1 - \frac{v^2}{c_0^2}}$$

(3)

This yields from Eq. (1), Eq. (2), and Eq. (3) the ratio of the familiar time dilation for the passage of time in the moving frame to be:

$$\Delta t_M = \frac{\Delta t_0}{\sqrt{1 - \frac{v^2}{c_0^2}}}$$

(4)

This expresses the fact that the moving electron's period is greater in the moving frame than for the stationary frame, in agreement with the Einstein relativistic time dilation for moving mass.

References

DT Froedge, "The Dirac Equation and the Two Photon Model of the Electron revised," April 2021, DOI:10.13140/RG.2.2.19095.70564, https://www.researchgate.net/publication/350922403.

DT Froedge, "The Electron as a Composition of Two Vacuum Polarization Confined Photons," April 2021, DOI: 10.13140/RG.2.2.18971.18722, https://www.researchgate.net/publication/350740864.

Hydrogen Fine Structure by the Feynman Flow Density Methods

D.T. Froedge

v92925

Abstract

This paper introduces the **Feynman flow density** as an alternative framework for describing the hydrogen atom and its fine structure. Rather than representing the electron by an abstract wavefunction, the flow density approach models the system as a **probability current of photons** bound within and around the electron and proton. These photon flows interact directly, giving rise to phenomena traditionally attributed to electric charge, but without requiring an external electric field concept[5].

By reformulating Feynman's path integral in terms of physical action flows instead of phase amplitudes, this method reproduces the full hydrogen spectral series and accounts for fine structure splittings. Calculated transition energies, derived from the interaction and collision of photon probability flows, are compared to high-resolution NIST data and found to agree within experimental uncertainties. The results suggest that a physically intuitive, photon-based mechanical picture can replace the Bohr orbit model while remaining consistent with the predictive power of Schrödinger and Dirac quantum mechanics.

Introduction

The hydrogen atom has long served as a test bed for quantum theory. Early models by **Bohr** and **Sommerfeld** offered a tangible mechanical picture of an electron in orbit, but modern quantum mechanics moved away from this visualization. Schrödinger's equation (1926) replaced orbits with standing wave solutions, giving a spatial probability density for the electron, while Dirac's relativistic formulation (1928) added electron spin and fine structure corrections. Although highly predictive, these formalisms no longer provide a **physical model** of hydrogen—an intuitive picture of what the electron and proton are doing.

This work explores an alternative perspective: the **Feynman flow density**. Instead of describing an electron as a pointlike charge with an associated wavefunction, we view it as a **bound state of photon flows**—dynamic, rotating photon currents both internal and external to the mass particle. The interaction of these flows between the electron and proton can reproduce all known spectral energy levels, including the fine structure, without introducing an independent electric field.

The aim of this paper is twofold:

1. To demonstrate that hydrogen's observed spectra can be derived by analyzing the **action flows** of the electron and proton using the flow-density formalism, rather than the standard wavefunction approach.
2. To provide a **physical and mechanical model** of the hydrogen atom that can conceptually replace the outdated Bohr picture while remaining compatible with the success of quantum mechanics.

The sections that follow present the core formulation of photon probability flows, derive the primary and fine spectral lines from these interactions, and the tables in the Appendix compare the computed results with precision data from the **NIST hydrogen spectral tables**. [7]

Main

This paper is a demonstration of the alternate view of the wavefunction [5] as a probability flow density of the photons that make up the electron,

The probability flow density is the an alternate approach to the Feynman integral over all paths, instead of integrating the wavefunctions phases over al paths the probability flow density integrates the physical action of the probability flow over all paths [5]. The photon probability density is a direct look at the same phenomena as the Feynman phase integral over all paths.

When the opposing flow probability of two opposite particles engages in head-on collisions, the colliding action flows produce photons that constitute the makeup of atomic spectral lines.

In interaction of atomic particle action flow was developed in "Details of Nuclear Particle Structure in Delta-c Mechanics". [4] The interaction between the probability flow density of the electron, and positron in hydrogen Illustrate that the interaction is a bit more complicated

From the flow action of proton in the hydrogen atom[] The action of the stationary electron action $S_e = \Re_0 m_e c_0 / \hbar$. The action in the hydrogen proton due to its most probable $n = 1$ paths is:

$$S_P = \int_0^{\Re_0} \frac{m_e c_0}{\hbar} dr = \frac{\Re_0 m_e c_0}{\hbar} = \frac{S_e}{\hbar} \rightarrow \frac{S_e}{\hbar g_A^2} = \alpha \qquad (1)$$

$\Re_0$ is the classical radius of the electron g_A^2 is the spin 2g factor accounting for the infinite Feynman path probabilities S_e is the electron state action.

For the electron this is the Path of the electron due to its most probable path around the proton has an action of:

$$S_E = \frac{S_e}{n_e 2\hbar g_A^2} = \frac{\alpha}{2}$$

Thus the well-known photon energy when $n_e = n_P \rightarrow n^2$ photons in the n=1 orbit is:

$$S_P S_e = \frac{S_e}{2n_e \hbar g_A^2} \frac{S_e}{n_P \hbar g_A^2} = \frac{1}{2} \frac{\alpha^2}{n^2} = 13.60569473 \text{ eV} \quad (2)$$

The result of this is not a surprise.

Sources of Spectral Lines

There are two sources of spectral lines in atomic hydrogen. One is the direct collision of opposing integral $\hbar$ action shown in Eq. (2).

$$\varepsilon = \frac{1}{2} \frac{\alpha^2}{n_X^2} \quad (3)$$

These are the maximum series line of the hydrogen atom with the Bohr interpretation of the change of energy from the free electron to the n_X line.

The other spectral lines are produced by the difference between in the energy of the spectral lines $\varepsilon = \varepsilon_2 - \varepsilon_2 = \hbar(v_2 - v_1)$. This is the beat frequency of two spectral photon streams, or the Rydberg energy levels:

$$\Delta\varepsilon = (\varepsilon_1 - \varepsilon_2) = h(v_1 - v_2) = \left(\frac{1}{2} \frac{\alpha^2}{n_X^2} - \frac{1}{2} \frac{\alpha^2}{n_Y^2} \right) = \frac{\alpha^2}{2} \left(\frac{1}{n_X^2} - \frac{1}{n_Y^2} \right) \quad (4)$$

These differences are responsible for the main Rydberg spectral lines, $(n_1, n_2, n_3 --)$.

Possible States of the
free Electron and Proton

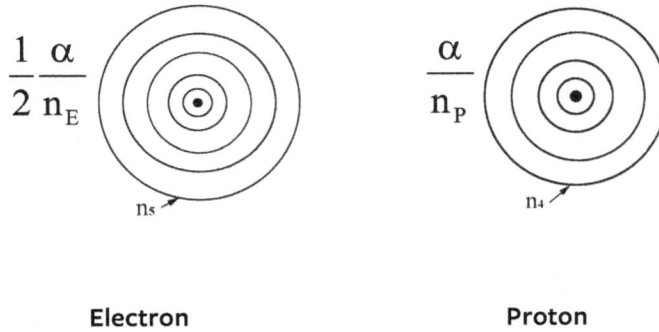

$$\frac{1}{2}\frac{\alpha}{n_E}$$

n_s

$$\frac{\alpha}{n_P}$$

n_4

Electron   **Proton**

To refine the calculations as the particles move together, the specific action of Feynman integral has to be taken into account for each level the effect of the photon position in a specific line is small but noticeable when high resolution values of the lines, such in the lamb offset is measured.

Proton Electron Probability
Flow Action

In summary the action of the electron around the proton the actions are:

$$S_P = \left(\frac{S_e}{n_P \hbar g_A}\right) = \frac{\alpha}{n_P} \qquad S_E = \left(\frac{S_e}{n_e \hbar g_A}\right) = \frac{\alpha}{2n_e} \qquad \frac{S_e}{\hbar} = \alpha g_A$$

(5)

The probability action of the electron and proton is the result of the rotation of the internal photons creating the particles. These action probabilities have a multiplicity of possible and probable values each of which is a multiple of the action quantum $\hbar$, thus:

$$f\left(n_P\right) = \frac{S_e}{\hbar n_P g_A^2} = \frac{\alpha}{1}\left(\frac{1}{1}, \frac{1}{2}, \frac{1}{3}, \frac{1}{4}, \frac{1}{5} - - -\right)$$

$$f\left(n_e\right) = \frac{S_e}{2\hbar n_e g_A^2} = \frac{\alpha}{2}\left(\frac{1}{1}, \frac{1}{2}, \frac{1}{3}, \frac{1}{4}, \frac{1}{5} - - -\right)$$

(6)

The probable action flow of the photons from the electron and proton in a bound condition are well known from the Bohr Theory.

From the probability action possibilities defined in Eq(6)., there is a multiplicity of other combinations of the flow of these photons that can generate the escaping photons. It is the energy contained in the electrons exterior probability flow $(13.6\,\text{eV})$ that can be radiated away as the particles bind.

Illustration

When the photon probability densities collide head-on, the energy of the transverse photons produced are equal to the product of the actions. The energy of the resulting photons for the for the case when $n_e = 3$ and $n_p = 3$ is shown in Fig. x .

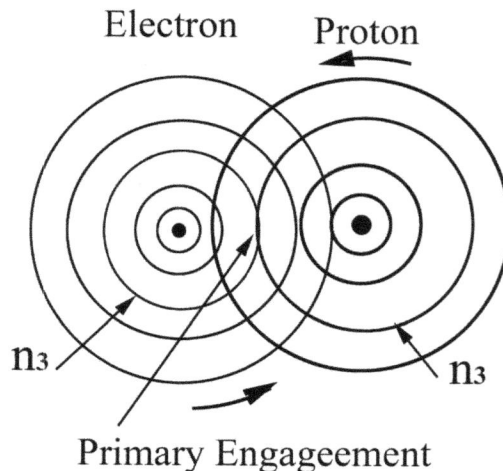

Primary Engageement

From the Bohr perspective this would be the spectral energy the first Paschen line, or defined by a free electron falling to the n_3 line.

Main Primary Spectral Radiation Lines

The momentum of two photons in collision cannot be conserved with an exchange of energy, but in the case of certain head-on collision this is not the case. The conditions can be satisfied, if there are two photons generated, and scatter laterally such that their energy is:

$$S_E = \frac{\alpha}{2n_e} \qquad S_P = \frac{\alpha}{n_P} \rightarrow \varepsilon_P = \frac{\alpha}{2n_e}\frac{\alpha}{n_P} \qquad (7)$$

See discussion in [x]

Photon Head-on-Collisions

Due to momentum considerations photons can only exchange energy under the certain circumstances of head on collisions. See discussion in [8]. The counter rotation of the flow of photons in a plane at 10E+20 cycles / sec, however provide the environment for such an exchange.

Photons in head-on collision of the probability flow are the source of most of the spectral lines and structure and provide the depletion of the photon flow levels as the particles engage.

$$\frac{\alpha}{n_P} > \frac{\alpha}{2n_e} \qquad\qquad\qquad \frac{\alpha}{n_P} < \frac{\alpha}{2n_e}$$

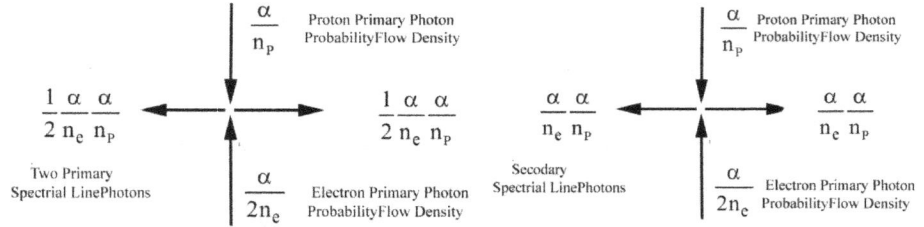

Fig. xx

This is the primary line for $n_P = n_e = 3$, and the energy is:

$$\varepsilon = \frac{1}{2} \frac{\alpha}{\times 3} \times \frac{\alpha}{3} = \frac{1}{2} \frac{\alpha^2}{3^2} = \frac{13.6eV}{9} \tag{8}$$

For the Bohr atom this would be photons coming from the maximum Paschen line.

The energy and frequency difference between the primary flow lines that produce the primary Bohr spectral lines, are:

$$\Delta\varepsilon = \hbar\Delta\nu = \frac{\alpha}{2n_1}\frac{\alpha}{n_1} - \frac{\alpha}{2n_2}\frac{\alpha}{n_2} \tag{9}$$

The interaction between the particles with opposing photon flow probability has the possibility of all the interactions between the lines as shown in Fig. x. The number of spectral lines and degeneracies possible are infinite, and thus form the background spectrum in photon emissions.

$$\alpha_P(n) \times \alpha_e(1,2,3,4,\cdots) \tag{10}$$

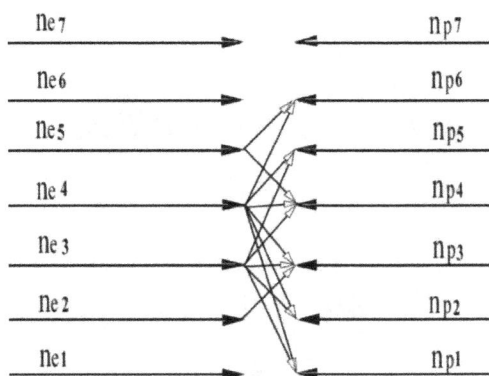

In Eq. (8), after arriving at this state position $3_P, 3_e$ the photons will be diminished by scattering the 1.51175 eV photons until the state is depleted, and the particles drop to the next level. The scattered photons can interact with the altered photons from other lines to create fine structure lines. Eventually the particles stop at the $n_e = 1$, $n_P = 1$ line with no relative rotation, and no exterior probability flow unable to escape.

Fine Structure

The creation of the primary spectra photons by the colliding main lines creates the possibility of a second generation of colliding photons within the hydrogen atom.

The generation of the fine lines has the same mechanics as the primary lines accept that the differential energies are smaller. The photons produced that generate the main spectral lines can collide with the photons of other such lines and produce photons of less energy.

Secondary Photon-Photon Emission from Head on Collisions

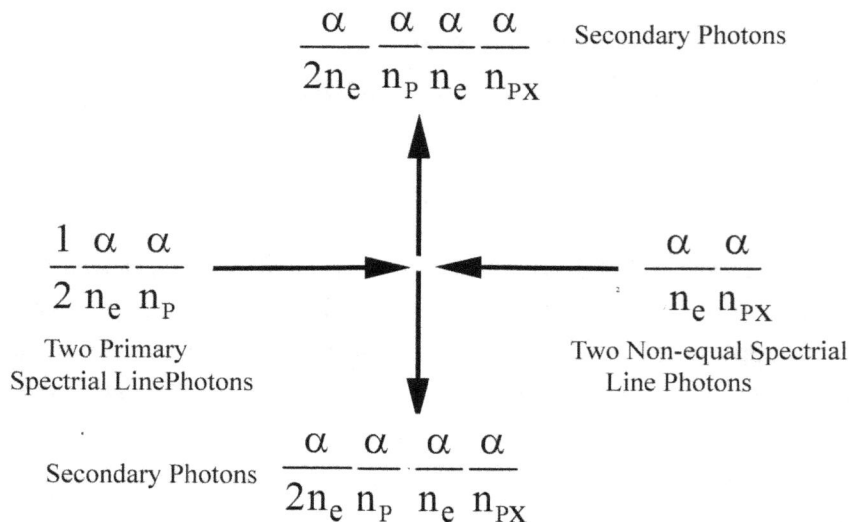

$$\frac{\alpha}{2n_e} \quad \frac{\alpha}{n_P} \quad \frac{\alpha}{n_e} \quad \frac{\alpha}{n_{PX}} \qquad \text{Secondary Photons}$$

$$\frac{1}{2}\frac{\alpha}{n_e}\frac{\alpha}{n_P} \longrightarrow \quad \longleftarrow \frac{\alpha}{n_e}\frac{\alpha}{n_{PX}}$$

Two Primary Spectrial LinePhotons

Two Non-equal Spectrial Line Photons

Secondary Photons $\quad \dfrac{\alpha}{2n_e} \; \dfrac{\alpha}{n_P} \cdot \dfrac{\alpha}{n_e} \; \dfrac{\alpha}{n_{PX}}$

Fig. Xx

As the difference in the in the primary spectral photons produce the primary spectral lines, the secondary fine structure lines are formed by the difference in the energy of the photons in the secondary collisions.

Secondary Line energy (Fine Structure)

$$\varepsilon\left(\Delta f\right) = \Delta v = \left(\frac{\alpha_e}{2n_{E1}} \frac{\alpha_P}{n_{P1}}\right)\left(\frac{\alpha_e}{n_{E2}} \frac{\alpha_P}{n_{P2}}\right) - \left(\frac{\alpha_e}{n_{E3}} \frac{\alpha_P}{n_{P3}}\right)\left(\frac{\alpha_e}{n_{E4}} \frac{\alpha_P}{n_{P4}}\right)$$

(11)

Orbital Action Contribution to the Electron and Proton orbits

Most of the calculated values presented in this paper are within the measured uncertainty of the spectral measurements, the differences of the NIST table for the fine lines has only 2 or 3 significant digits. The action of the Electron Proton Compton orbits of the particles at the multiple n levels contribute, and thus for accurate calculations, they need to be evaluated.

The main action can be determined by the increase of the orbits around the center of mass. As the electron orbits the proton, the action of the proton is altered by

$$S_P\left(proton\right) \rightarrow \frac{S_e}{n_P \hbar}\left(1 + \frac{m_e}{m_P}\right)$$

(12)

The proton is relatively stationary thus the action of its electron probability flow is equivalent to that of the stationary electron S_e.

$$S_P = \left(1 + \frac{m_e}{m_P}\right) \qquad = \qquad 1.00054461603$$

(13)

The NIST spectral offset in spectrum due to the ionization correction factor is 1.000533776, and has the same effect on the absolute value of spectral lines as does the orbital mass adjustment of Eq.(13),

This affects the energy of the fine spectral lines such that:

$$\varepsilon_F = \left[\left(\frac{\alpha_e}{2n_L} \frac{\alpha_P}{n_L} \right) \left(\frac{\alpha_e}{n_L} \frac{\alpha_P}{n_X} \right) - \left(\frac{\alpha_e}{2n_L} \frac{\alpha_P}{n_L} \right) \left(\frac{\alpha_e}{n_L} \frac{\alpha_P}{n_Y} \right) \right] \times \left(1 + \frac{2m_e}{m_P} \right)$$

(14)

The differential effect is small, but it is included in the calculated Appendix data.

Eq.(14), can be compared with the Bohr Summerfield fine structure term from the expression:

$$\varepsilon_S = \left[1 - \frac{\alpha^2}{2n^2} - \frac{\alpha^4}{2n^4} \left(\frac{n}{l+1/2} - \frac{3}{4} \right) \right]$$

(15)

Lamb Shift

There are adjustments to the action due to the position of the orbits in in the various states. The determination of this changes are small and its not clear how it would be done. These levels are for the most part below the error in the measured values, and only the value of the adjustment defined in Eq.(13), is included in the data tables provided in this paper.

For one value however, the 1057 Lamb line, the exact action can be back calculated.

For the 2 p2 , 2 s2 Lamb line The frequency is 1057.833 (4) MHz, and the fine structurer action has an energy od 4.3748457e-6 eV. This yields a more accurate value of the action for the $n_e = n_p = 2$ line.

It yields:

$$\rightarrow \left(S_E S_P \right)^2 m_R^2 = \left(S_E S_P \right)^2 \left(1 + \frac{2m_e}{m_P} + \frac{1}{2\pi} \frac{m_e}{m_P} \right) \qquad (16)$$

The additional third term in the action (8.6678e-5), results in the value of the calculated value to be within the experimental error.

Calculated vs Measured Spectral Line Energy

The following calculations of the fine structure are provided in the tables of Appendix I, are generated by Eq.(14). The first entry in the tables are the primary line number, the second is the difference between the 2p line and the other fine structure lines shown in the NIST table in. The third is the calculated comparative difference between the photon energy of the two lines from Eq.(14). The 4-8 are the photon energy of the n p2 line, and 10-13 are the value and n numbers of the photon. Most of the 2 p2 lines are slightly different

   The 2p lines are primarily the product of the n electron proton n^2 line produced mostly by the collision of the primary line and the n, 1 line with another similar photon that the collision of the same n^2 line with n_X

$$\varepsilon_P = \left(\frac{\alpha}{2 \times n} \frac{\alpha}{n} \right) \left(\frac{\alpha}{n} \frac{\alpha}{1} \right) \qquad \text{and} \qquad \varepsilon_P = \left(\frac{\alpha}{2 \times n} \frac{\alpha}{n} \right) \left(\frac{\alpha}{n} \frac{\alpha}{n_X} \right)$$

$$(17)$$

Appendix II is the same with just the $2p - n\,2s$ lines.

   Appendix III is the NIST spectral data with the inclusion of the difference between the 2p and the 2s or the lamb lines of each primary line

Conclusion

This work has shown that the **Feynman flow density** formulation provides a physically interpretable alternative to the conventional wavefunction description of the hydrogen atom. By replacing abstract phase amplitudes with a concrete picture of photon probability flow surrounding the bound electron and proton, the model recovers the principal spectral lines and their fine structure without invoking an external electric field or the classical Bohr orbit picture. Instead, the observed hydrogen spectrum emerges naturally from the head-on interaction of photon flows: primary collisions produce the main Balmer, Paschen, and higher-series photons, while secondary photon–photon interactions generate the observed fine structure splittings.

The approach connects directly to the Feynman path integral but reframes it in terms of action flow densities—quantities that can be assigned to the rotating, self-bound photons forming the electron and proton. By evaluating the action at each bound state and incorporating small orbital corrections, the calculated line positions agree with high-resolution NIST data to within their stated uncertainties, including subtle offsets such as those associated with the Lamb shift.

Beyond matching known spectra, the flow-density view restores a mechanical picture of the hydrogen atom that modern quantum mechanics has largely abandoned. It offers a direct visualization of how internal photon dynamics can account for mass, charge-like interaction, and radiative transitions, while remaining compatible with the successful predictions of Dirac theory and quantum electrodynamics.

Future work may extend these methods to more complex atoms and to nuclear systems where multiple probability flows overlap. Refining the small action terms responsible for ultra-fine corrections could deepen the link between this physically motivated picture and standard QED treatments. Most importantly, the results demonstrate that the hydrogen fine structure need not rely solely on abstract wave equations; it can also be understood through the photon flow mechanics underlying Delta-c theory.

References:

DT Froedge, The physics of Delta-c Mechanics, ISBN-13 979-8218347178 (Feb. 14, 24) https://www.amazon.com/Physics-Delta-C-Mechanics-Approach-Particle/dp/B0CVZ8CNYQ

1. DT Froedge, The Concepts and Principles Behind of Delta-c Mechanics, August 2024, DOI:10.13140/RG.2.2.31630.37445, https://www.researchgate.net/publication/382850589

2. DT Froedge, The Connection between Electric Charge, Gravitation, and the Feynman Sum over All Histories View of Quantum Electrodynamics, April 2020 Conference: APS April. 18-21, 2020 Washington DC., https://absuploads.aps.org/presentation.cfm?pid=18355 https://www.researchgate.net/publication/341310206

3. DT Froedge, Calculated Atomic Masses in Delta-c Mechanics January 2025 DOI: 10.13140/RG.2.2.30292.51842 https://www.researchgate.net/publication/388219158

4. DT Froedge, Structure of Elementary Nuclear Particles in in Delta-c Mechanics November 2024   DOI: 10.13140/RG.2.2.27181.70884 https://www.researchgate.net/publication/385782943

5. DT Froedge, Feynman Flow Density Alternative to Wavefunctions, June 25 DOI:10.13140/RG.2.2.11759.14242, https://www.researchgate.net/publication/392330748

6. DT Froedge, Feynman Vertex Functions, June 2025.DOI:10.13140/RG.2.2.35247.24487

7. NIST Hydrogen Database, https://physics.nist.gov/cgi-bin/ASD/energy1.pl?de=0&spectrum=hydrogen&submit=Retrieve+Data&units=1&format=0&output=0&page_size=15&multiplet_ordered=0&conf_

out=on&term_out=on&level_out=on&unc_out=1&j_out=on&lande_out=on&perc_out=on&biblio=on&temp=

8. G. Sandhu, I. Dhindsa, Cosmological redshift caused by head-on collisions with CMB photons, July 2024. DOI: 10.13140/RG.2.2.30466.49608, https://www.researchgate.net/publication/38233544

Appendix I

Fine structure line calculations

Hydrogen Fine Structure   **Lines 2-12**

$$\varepsilon_F = \left[\left(\frac{\alpha}{2n_L}\,\frac{\alpha_P}{n_L}\right)\left(\frac{\alpha}{n_L}\,\frac{\alpha}{n_X}\right) - \left(\frac{\alpha}{2n_L}\,\frac{\alpha}{n_L}\right)\left(\frac{\alpha}{n_X}\,\frac{\alpha}{n_Y}\right)\right]\left(1+\frac{2m_e}{m_P}\right)$$

Line #	NIST Hydrogen Fine Line Delta		$\left(\frac{\alpha}{2n_1}\frac{\alpha}{n_2}\frac{\alpha}{2n_3}\frac{\alpha}{n_4}\right)$	n_L	n_L	n_L	n_X	$\left(\frac{\alpha}{2n_1}\frac{\alpha}{n_2}\frac{\alpha}{2n_3}\frac{\alpha}{n_y}\right)$	n_L	n_L	n_L	n_Y
2	4.540000E-05	4.533190E-05	9.066380E-05	2	2	2	1	4.533190E-05	2	2	2	2
2	4.374890E-06	4.374130E-06	7.555320E-06	2	2	2	12	3.181188E-06	2	2	3	19
2	2.527490E-05	2.518440E-06	7.555320E-06	2	2	2	12	5.036880E-06	2	2	2	18
2	1.571460E-05	4.506680E-06	5.036880E-06	2	2	3	12	9.543563E-06	2	2	1	19
3	1.789910E-05	1.790890E-05	2.686340E-05	3	3	3	1	8.950000E-06	3	3	3	3
3	1.341920E-05	1.343170E-05	2.686340E-05	3	3	3	1	1.340000E-05	3	3	3	2
4	7.551680E-06	7.555320E-06	1.133300E-05	4	4	4	1	3.780000E-06	4	4	4	3
4	5.670420E-06	5.666490E-06	1.133300E-05	4	4	4	1	5.670000E-06	4	4	4	2
4	5.662080E-06	5.666490E-06	1.133300E-05	4	4	4	1	5.670000E-06	4	4	4	2
5	3.860000E-06	3.868320E-06	5.802490E-06	5	5	5	1	1.930000E-06	5	5	5	3
5	2.904000E-06	2.901240E-06	5.802490E-06	5	5	5	1	2.900000E-06	5	5	5	2
5	4.347300E-06	4.351860E-06	5.802490E-06	5	5	5	1	1.450000E-06	5	5	5	4
5	4.637630E-06	4.641990E-06	5.802490E-06	5	5	5	1	1.160000E-06	5	5	5	5

6	2.796290E-06	2.798270E-06	3.357920E-06	6	6	6	1	5.600000E-07	6	6	6	6
6	2.684280E-06	2.686340E-06	3.357920E-06	6	6	6	1	6.720000E-07	6	6	6	5
6	2.516590E-06	2.518440E-06	3.357920E-06	6	6	6	1	8.390000E-07	6	6	6	4
6	2.237360E-06	2.238610E-06	3.357920E-06	6	6	6	1	1.120000E-06	6	6	6	3
6	1.677430E-06	1.678960E-06	3.357920E-06	6	6	6	1	1.680000E-06	6	6	6	2
7	1.030000E-07	1.057300E-06	2.114610E-06	7	7	7	1	1.060000E-06	7	7	7	2
8	7.080000E-07	7.083110E-07	1.416620E-06	8	8	8	1	7.083110E-07	8	8	8	2
9	4.960000E-07	4.974700E-07	9.949390E-07	9	9	9	1	4.970000E-07	9	9	9	2
9	4.800000E-08	4.974700E-07	9.949390E-07	9	9	9	1	4.970000E-07	9	9	9	2
	6.400000E-07	6.632930E-07	9.949390E-07	9	9	9	1	3.316460E-07	9	9	9	3
10	4.820700E-07	4.835410E-07	7.253110E-07	10	10	10	1	2.420000E-07	10	10	10	3
10	3.630000E-07	3.626550E-07	7.253110E-07	10	10	10	1	3.630000E-07	10	10	10	2
10	3.610800E-07	3.626550E-07	7.253110E-07	10	10	10	1	3.630000E-07	10	10	10	2
10	3.600000E-08	3.626550E-07	7.253110E-07	10	10	10	1	3.630000E-07	10	10	10	2
11	3.620000E-07	3.632910E-07	5.449370E-07	11	11	11	1	1.820000E-07	11	11	11	3
11	2.730000E-07	2.724680E-07	5.449370E-07	11	11	11	1	2.720000E-07	11	11	11	2
11	2.710000E-07	2.724680E-07	5.449370E-07	11	11	11	1	2.720000E-07	11	11	11	2
11	2.700000E-08	2.724680E-07	5.449370E-07	11	11	11	1	2.720000E-07	11	11	11	2
12	2.778200E-07	2.798270E-07	4.197400E-07	12	12	12	1	1.400000E-07	12	12	12	3
12	2.720000E-07	2.798270E-07	4.197400E-07	12	12	12	1	1.400000E-07	12	12	12	3
12	2.090000E-07	2.098700E-07	4.197400E-07	12	12	12	1	2.100000E-07	12	12	12	2
12	2.078100E-07	2.098700E-07	4.197400E-07	12	0	0	1	2.100000E-07	12	12	12	2

Appendix II

Hydrogen Fine Structure, Lamb Lines

n 2p.- n 2s Lines

$$\varepsilon = \left[\left(\frac{\alpha}{2n_L}\frac{\alpha}{n_L}\right)\left(\frac{\alpha}{n_L}\frac{\alpha}{n_{12}}\right) - \left(\frac{\alpha}{2n_L}\frac{\alpha}{n_L}\right)\left(\frac{\alpha}{n_X}\frac{\alpha}{n_Y}\right)\right] \times \left(1 + \frac{2m_e}{m_P}\right)$$

Line #	NIST Hydrogen Fine Line Delta	↓	$\left(\frac{\alpha}{2n_1}\frac{\alpha}{n_2}\frac{\alpha}{2n_3}\frac{\alpha}{n_4}\right)$	n_L	n_L	n_L	n_{12}	$\left(\frac{\alpha}{2n_1}\frac{\alpha}{n_2}\frac{\alpha}{2n_3}\frac{\alpha}{n_Y}\right)$	n_L	n_L	n_X	n_Y
2	4.37485E-06	4.37413E-06	7.55532E-06	2	2	2	12	3.18119E-06	2	2	6	19
3	1.30191E-06	1.29604E-06	2.23861E-06	3	3	3	12	9.42574E-07	3	3	9	19
4	5.50310E-07	5.46767E-07	9.44415E-07	4	4	4	12	3.97648E-07	4	4	12	19
5	2.82000E-07	2.79945E-07	4.83541E-07	5	5	5	12	2.03596E-07	5	5	15	19
6	1.63000E-07	1.62005E-07	2.79827E-07	6	6	6	12	1.17822E-07	6	6	18	19
7	1.03000E-07	1.02021E-07	1.76217E-07	7	7	7	12	7.41968E-08	7	7	21	19
8	6.90000E-08	6.83458E-08	1.18052E-07	8	8	8	12	4.97061E-08	8	8	24	19
9	4.80000E-08	4.80015E-08	8.29116E-08	9	9	9	12	3.49102E-08	9	9	27	19
10	3.60000E-08	3.49931E-08	6.04426E-08	10	10	10	12	2.54495E-08	10	10	30	19
11	2.70000E-08	2.62908E-08	4.54114E-08	11	11	11	12	1.91206E-08	11	11	33	19
12	2.00000E-08	2.02506E-08	3.49783E-08	12	12	12	12	1.47277E-08	12	12	36	19

Note- in this table action for n 2p.- n 2s s set as In Eq.(16), to illustrate the effect on the lamb shift.

$$\left(1 + \frac{2m_e}{m_P} + \frac{1}{2\pi}\frac{m_e}{m_P}\right) \tag{18}$$

For the other lines the inclusion is not significant.

Appendix III

NIST Atomic Spectra Database Levels Data [7]

With the differences between the 2p lines and the

The following data is from the NIST database with the difference between the 2P and the other fine lines (Column 6) and the reference line, column 7

H I   106 Levels Found
Z = 1, H isoelectronic sequence

1	2	3	4	5	6	7
					Delta with 2P	
Config	Term	J	Level (eV)	Uncertainty	- L-2p	
1s	2S	1/2	0	1.20E-13		1
2p	2P°	1/2	10.19880615	4.00E-11		
		3/2	10.19885151	5.00E-11	4.53643500E-05	2
2s	2S	1/2	10.19881053	1.20E-13	4.37490810E-06	2
2			10.1988358	1.20E-06	2.96497600E-05	2
3p	2P°	1/2	12.08749366	9.00E-10		3
		3/2	12.0875071	9.00E-10	1.34413000E-05	3
3s	2S	1/2	12.08749496	9.00E-10	1.30200000E-06	3
3			12.0875052	6.00E-07	1.15409000E-05	3
3d	2D	3/2	12.08750708	2.50E-09	1.34192000E-05	3
		5/2	12.08751156	9.00E-10	1.78991000E-05	3
4p	2P°	1/2	12.74853245	6.00E-11		4
		3/2	12.74853812	4.00E-11	5.67042000E-06	4
4s	2S	1/2	12.748533	4.00E-11	5.50310000E-07	4
4d	2D	3/2	12.74853811	1.00E-09	5.66208000E-06	4
		5/2	12.74854	1.00E-10	7.55121000E-06	4
4			12.7485393	4.00E-07	6.85368000E-06	4
4f	2F°	5/2	12.74854	6.00E-09	7.55168000E-06	4
		7/2	12.74854094	1.60E-09	8.49398000E-06	4
5p	2P°	1/2	13.05449818	5.00E-09		

		3/2	13.05450109	5.00E-09	2.90400000E-06	5
5s	2S	1/2	13.05449846	5.00E-09	2.82000000E-07	5
5d	2D	3/2	13.05450107	3.00E-09	2.89200000E-06	5
		5/2	13.05450204	3.00E-09	3.86000000E-06	5
5			13.0545017	5.00E-07	3.51800000E-06	5
5f	2F°	5/2	-13.05450205	1.00E-11	3.86433600E-06	5
		7/2	13.05450253	6.00E-09		
5g	2G	7/2	-13.05450253	1.00E-11	4.34730300E-06	5
		9/2	-13.05450282	1.00E-11	4.63763300E-06	5
6p	2P°	1/2	13.22070146	1.20E-10		6
		3/2	13.22070314	1.20E-10	1.68016000E-06	6
6s	2S	1/2	13.22070163	9.00E-11	1.63340001E-07	6
6d	2D	3/2	13.22070314	1.60E-10	1.67743000E-06	6
		5/2	13.2207037	5.00E-11	2.23736000E-06	6
6f	2F°	5/2	-13.2207037	1.00E-11	2.23710100E-06	6
		7/2	-13.22070398	1.00E-11	2.51712300E-06	6
6			13.22070389	2.00E-07	2.42802000E-06	6
6g	2G	7/2	-13.22070398	1.00E-11	2.51659400E-06	6
		9/2	-13.22070415	1.00E-11		
6h	2H°	9/2	-13.22070415	1.00E-11	2.68427800E-06	6
		11/2	-13.22070426	1.00E-11	2.79629200E-06	6
7p	2P°	1/2	13.32091665	5.00E-09		
		3/2	13.3209177	5.00E-09	1.05700000E-06	7
7s	2S	1/2	13.32091675	5.00E-09	1.03000000E-07	7
7d	2D	3/2	13.3209177	3.00E-09	1.05600000E-06	7
		5/2	13.32091806	3.00E-09		
7			13.32091817	1.10E-07	1.52300000E-06	7
8p	2P°	1/2	13.38596008	7.00E-11		
		3/2	13.38596079	7.00E-11	7.08820000E-07	8
8s	2S	1/2	13.38596015	4.00E-11	6.89600004E-08	8
8d	2D	3/2	13.38596079	4.00E-11	7.07670001E-07	8
		5/2	13.38596102	2.50E-11	9.43949001E-07	8
8			13.38596103	9.00E-08	9.51310001E-07	8

9p	2P°	1/2	13.4305536	5.00E-09		9
		3/2	13.4305541	5.00E-09	4.98000000E-07	9
9s	2S	1/2	13.43055365	5.00E-09	4.80000004E-08	9
9d	2D	3/2	13.4305541	3.00E-09	4.96000000E-07	9
		5/2	13.43055426	3.00E-09	6.62000000E-07	9
9			13.43055424	9.00E-08	6.40000000E-07	9
10p	2P°	1/2	13.46245106	5.00E-09	3.60000012E-08	10
		3/2	13.46245142	5.00E-09	3.63000002E-07	10
10s	2S	1/2	13.46245109	5.00E-09	3.60000012E-08	10
10d	2D	3/2	13.46245142	1.60E-10	3.61080001E-07	10
		5/2	13.46245154	1.60E-10	4.82070000E-07	10
10			13.46245155	1.00E-07	4.92000002E-07	10
11p	2P°	1/2	13.48605155	5.00E-09		11
		3/2	13.48605183	5.00E-09	2.73000001E-07	11
11s	2S	1/2	13.48605158	5.00E-09	2.70000005E-08	11
11d	2D	3/2	13.48605183	3.00E-09	2.71000001E-07	11
		5/2	13.48605192	3.00E-09	3.62000000E-07	11
11			13.48605194	7.00E-08	3.86000000E-07	11
12p	2P°	1/2	13.50400166	5.00E-09		
		3/2	13.50400187	5.00E-09	2.09000000E-07	12
12s	2S	1/2	13.50400168	5.00E-09	1.99999999E-08	12
12d	2D	3/2	13.50400187	4.00E-11	2.07810000E-07	12
		5/2	13.50400194	3.00E-11	2.77820000E-07	12
12			13.50400193	6.00E-08	2.71999999E-07	12
	Limit	---	-13.5984346	1.20E-11		

Index

rest mass, 27, 42, 55, 72, 76, 77, 91–92, 99, 118–23. *See also* mass values for specific particles

Rydberg integers, 7, 24, 34, 41, 48–49, 58, 61, 94–95, 98, 101

Rydberg levels, 13, 25, 37, 59, 97, 162

Rydberg states, 5, 7, 8, 11, 32, 33–34, 36, 57, 87, 94, 95, 103

Sandhu, G., 174

scalar energy, 5, 7, 30

Schrödinger equation, 27, 117, 160

Schwinger, J., 21n17

Shapiro, I., 148n17

spectral energy, 160, 165, 171

spectral lines, 161, 162, 165, 166–71

spin, 33–34, 35, 47, 128, 154, 160, 161

state energy, 5, 7, 26, 62, 97

state functions rationalization, 48–49

Stone, Melvin L., 148n17

tauons

      calculated mass of, 40, 46, 64–66, 106, 109

      structure of, 127

      vertex functions of, 76–77

      vertex vs. ejection energy, 78

Thorne, Kip S., 17, 21n15, 143, 148n13, 148n15

time dilation, 155–57

top quarks, 40, 42, 64–65, 67, 110

transition energies, 159

vacuum density, 15–17, 129, 131, 142–44, 146

Valev, D., 148n16

vertex functions. *See also* kernels

      overview, ix, 27–28

      atomic vs. nuclear kernels, 8–10, 31, 33–35, 36, 42, 59–60, 62–63, 86–87

      and calculated mass values for specific particles, 39–42, 54, 55, 62–64

      and conjugate particles, 29–30

      electron neutrino vertex function, 73–74

      Feynman flow function, 32

      Feynman flow probability density, 11, 12, 29

      illustrations of, 86–88

      multiple kernel nuclear vertex function, 36–37

      muon vertex function, 14, 38, 85, 86

      neutrino vertex function, 72–78, 88

      neutron vertex function, 63, 88

      nuclear vertex function, 10

      pion vertex function, 63, 75–76, 77, 88

      proton vertex function, 72–73

      specific particle illustrations, 37–38

      and state functions rationalization, 49

      tauon neutrino vertex function, 76–77